GEOLOGICAL PROBLEM SOLVING WITH LOTUS 1-2-3 FOR EXPLORATION AND MINING GEOLOGY

COMPUTER METHODS IN THE GEOSCIENCES

Daniel F. Merriam, Series Editor

Previously published by Van Nostrand Reinhold Co. Inc.

Computer Applications in Petroleum Geology
Joseph E. Robinson
Graphic Display of Two- and Three-Dimensional Markov Computer Models in Geology
Cunshan Lin and John W. Harbaugh
Image Processing of Geological Data
Andrea G. Fabbri
Contouring Geologic Surfaces with the Computer
Thomas A. Jones, David E. Hamilton, and Carlton R. Johnson
Exploration-Geochemical Data Analysis with the IBM PC
George S. Koch, Jr. (with programs on diskettes)
Geostatistics and Petroleum Geology
M. E. Hohn
Simulating Clastic Sedimentation
Daniel M. Tetzlaff and John W. Harbaugh

* Orders to: Van Nostrand Reinhold Co. Inc, 7625, Empire Drive, Florence, KY 41042, USA

Related Pergamon Publications

Books

GREEN
Computer Applications for Geologic Exploration: Resource and Hazards

HANLEY & MERRIAM (Editors)
Microcomputer Applications in Geology II

Journals

Computers & Geosciences

Computer Languages

Information Processing & Management

International Journal of Rock Mechanics and Mining Sciences (& Geomechanics Abstracts)

Minerals and Engineering

Full details of all Pergamon publications/free specimen copy of any Pergamon journal available on request from your nearest Pergamon office.

GEOLOGICAL PROBLEM SOLVING WITH LOTUS 1-2-3 FOR EXPLORATION AND MINING GEOLOGY

(with programs on diskette)

GEORGE S. KOCH, JR., University of Georgia

PERGAMON PRESS

Member of Maxwell Macmillan Pergamon Publishing Corporation

OXFORD · NEW YORK · BEIJING · FRANKFURT

SÃO PAULO · SYDNEY · TOKYO · TORONTO

U.K.	Pergamon Press plc, Headington Hill Hall, Oxford OX3 0BW, England
U.S.A.	Pergamon Press, Inc., Maxwell House, Fairview Park, Elmsford, New York 10523, U.S.A.
PEOPLE'S REPUBLIC OF CHINA	Pergamon Press, Room 4037, Qianmen Hotel, Beijing, People's Republic of China
FEDERAL REPUBLIC OF GERMANY	Pergamon Press GmbH, Hammerweg 6, D-6242 Kronberg, Federal Republic of Germany
BRAZIL	Pergamon Editora Ltda, Rua Eça de Queiros, 346, CEP 04011, Paraiso, São Paulo, Brazil
AUSTRALIA	Pergamon Press Australia Pty Ltd., P.O. Box 544, Potts Point, N.S.W. 2011, Australia
JAPAN	Pergamon Press, 5th Floor, Matsuoka Central Building, 1-7-1 Nishishinjuku, Shinjuku-ku, Tokyo 160, Japan
CANADA	Pergamon Press Canada Ltd., Suite No. 271, 253 College Street, Toronto, Ontario, Canada M5T 1R5

First edition 1990

Library of Congress Cataloging-in-Publication Data

Koch, George S.
Geological problem solving with LOTUS 1-2-3 for exploration and mining geology (with programs on diskette)/George S. Koch, Jr.
p. cm.—(Computer methods in the geosciences: 8)
Includes bibliographical references and index.
1. Prospecting—Statistical methods—Data processing.
2. Mining geology—Statistical methods—Data processing.
3. LOTUS 1-2-3 (Computer program) I. Title. II. Series.
TN270.K59 1990 622 s—dc20 [622'.1]
90-7335

British Library Cataloguing in Publication Data

Koch, George S. (George Schneider), *1926–*
Geological problem solving with Lotus 1-2-3 for exploration and mining geology (with programes on diskette).
1. Geology. Applications of computer systems
I. Title
551.0285

ISBN 0-08-036941-3 Hard cover
ISBN 0-08-040281-X Flexicover

Printed in Great Britain by BPCC Wheatons Ltd, Exeter

Contents

List of Worksheets

Series Editor's Foreword

This contribution to the Computer Methods in the Geosciences (CMG) series by George Koch, Jr., is his second - the first being, "Exploration-Geochemical Data Analysis with the IBM PC." As with the first book, this one is another practical guide. This work is the direct result of Koch's 35 years experience in the exploration for and exploitation of mineral deposits. Coauthor with R.F. Link of the successful book on "Statistical Analysis of Geological Data" (v. 1, 1970/v. 2, 1971, John Wiley & Sons), it brings a wealth of experience to the material presented in this volume.

The reader is taken through the book step-by-step with detailed instructions on just how to accomplish each step. Subjects range through statistics, such as confidence intervals, fitting a straightline, analysis of variance, probability, and exploration models to exploration geochemical data, from economic considerations such as compound interest, depreciation and depletion, and discounted cash flow to blocking ore from workings and drillhole data, ore concentration and smelter settlement to Peters's model for mineral-property evaluation. The object of the book is to teach the user how Lotus 1-2-3 can help solve the many numerical problems facing the practitioner today.

Spreadsheets are being used more and more as earth-science users learn of the many applications to which they can be made. They are versatile and easy to use. Koch has adeptly put together the 1-2-3 programs as well as the text explaining their use. Examples are taken from his own work as well as from the literature and range from Tennessee zinc to Nevada placer gold deposits. Intended to be used by both the professional and student, this book can serve as a text/lab manual or a reference book. It should be noted too, that the material is so structured that it could be used in other geological fields, including petroleum geology.

This volume is the eighth in the series initiated in 1982. First published by Hutchinson Ross Publishing Company (Stroudsburg, PA) and later Van Nostrand Reinhold (New York), the series is being continued by Pergamon Press. The series (1) promotes the subject of geomathematics by making available material in plain English; (2) introduces the reader to the subject and where to get more information; and (3) keeps the geological public informed of latest developments in a fast-moving field. As stated in the first volume of the series it is ".....designed to be self-contained and open-ended." The same high quality is promised by the publisher and this contribution by George Koch certainly fulfills that promise.

Other titles which are in preparation include:

- Computer Applications for Geologic Exploration: Resource and Hazards, by William Green.

- Simulating Sedimentary Transport by Waves, by Paul A. Martinez and John W. Harbaugh.

DANIEL F. MERRIAM

Preface

Geologists working in the mineral industries are accustomed to solving many numerical problems, some concerned with exploring for deposits, and others with evaluation of deposits once they are located. Required is a knowledge of varied techniques, most of which may be classified as statistical or financial. The software package 1-2-3 (TM) of the Lotus Development Corporation provides a powerful new tool for implementing these techniques. The purpose of this book is to present effective 1-2-3 problem-solving methods for geology as applied in the mineral industries.

This book provides 1-2-3 programs (also named worksheets) together with a brief text that explains them. The statistical and financial principles, not explained in this book, may be presented in books for which this book may be a companion work.

This book is for both professionals and students. Professionals will determine it suitable for self-study. It also can be used as a laboratory manual for students.

Using the floppy diskette supplied, you will be able to solve problems at once, following the brief instructions. For effective learning, I have structured the ideas and worksheets from simple to complex. In order to make it easy for you to modify the worksheets if you wish, I have made them simple and have defined and labeled the variables clearly. Advanced features of 1-2-3 are used only when absolutely necessary.

1-2-3 is ideally suited for our purpose because it combines a spreadsheet with mathematical and statistical analysis and graphical displays. The term *spreadsheet* implies in accounting "the traditional financial modeling tools: the accountant's columnar pad, pencil, and calculator" (Ewing, D.P., and others, 1987, p. 12). In geology, the spreadsheet becomes an orthogonal x,y grid on which to drillholes, take soil samples, or lay out a mining plan. And, in mathematical/statistical terms, the spreadsheet becomes a matrix of data or calculated terms.

Therefore, with 1-2-3 your only investment in time (or, overhead in computer language) is learning its rudiments, rather than having to learn several systems, perhaps one for statistics, one for databases, and yet another one for financial calculations. And, because 1-2-3 is used widely, many books are available if you need them. Help also is available from the Lotus Development Corp. Moreover, because of the widespread acceptance of 1-2-3, we can anticipate that the system will be maintained in the future. This book's programs should be compatible with future releases of updated versions of 1-2-3.

I have learned programming in 1-2-3 is highly rewarding in contrast to programming in a high-level language such as FORTRAN or BASIC. Programming in 1-2-3 is to programming in BASIC or FORTRAN as building with premade window and door units is to building with only dimension lumber and glass. Correcting any bugs that may exist in this book or in other programs that you may write will be less of a problem than in high-level languages because of 1-2-3's structure and organizational support.

Although the programs in this book are written for problems about metallic ore deposits, they also are pertinent for industrial minerals and are relevant for problem solving in other geological specialties including petroleum geology.

The example problems come from many sources. Some are from the excellent 1987 book "Exploration and Mining Geology," by W. Peters

(John Wiley & Sons, New York); in fact, I have included most of Peters's numerical examples; others were taught me by H.E. McKinstry. A number are from my work, published and unpublished.

T.J. Bornhorst, J.T. Hanley, D. Papacharalampos, W.C. Peters, and T.B. Thompson reviewed various drafts of this book and made many helpful suggestions. Of course, I am responsible for any mistakes and shortcomings.

GEORGE S. KOCH, JR.

Disclaimer

User Assistance and Information

Any problems, comments, or suggestions regarding these problems should be directed to Dr. George S. Koch, Jr., Department of Geology, University of Georgia, Athens, GA 30602.

CHAPTER 1

Introduction

This book contains Lotus 1-2-3 worksheets (on a floppy diskette) and a brief explanatory text. In addition, you need to have 1-2-3 (version 2.01 or later) available.

I assume that you have a basic understanding of 1-2-3. If you are unfamiliar with the system, the "Introduction to 1-2-3" of the Lotus Development Corporation will be helpful. The Corporation's 1-2-3 manuals are useful for reference. An excellent introduction to 1-2-3 in general is by Ewing and others (1987), and a book on statistics with 1-2-3 has been written by Kilpatrick (1987).

I assume that you are familiar with the elementary concepts and standard notation of mathematical statistics, available in a geological framework in books by Koch and Link (1970, 1971), Davis (1986), or in standard statistical textbooks. Also assumed is familiarity with basic methods of mineral-deposit evaluation, as developed by Peters (1987) in a geologically oriented book or by Stermole and Stermole (1987) in a book on economic evaluation.

The floppy diskette contains 1-2-3 worksheets, which have the property that the model, consisting of the formulae, is not separated from the data, which consist of the values that change from one problem to another. This situation is unlike that in higher level languages (for example, FORTRAN or BASIC), in which the data and formulae are clearly separated. I provide worksheets for representative geological and mining situations. To use the worksheets for your data, simply replace the example data with your data, altering the numbers of rows and columns if necessary. Alternatively, remove all of the data, to create dataless worksheets, named *templates*, in 1-2-3 jargon. (Be careful to remove only data, not formulae; you can identify data cells by inspection on the screen, or by printing out the contents of all cells, using the /Print/ Printer/Options /Other/Cell-Formula command.)

The rest of this chapter tells you how to get started using the worksheets, provides some other introductory material, and ends with a general conclusion for this book.

HOW TO START

Many of us prefer to see something on the screen before reading text. If you belong to this group, here is how to start.

You need an IBM Personal Computer, with enough memory to run 1-2-3, and preferably with a graphics card, or a compatible computer made by another vendor. Desirable but not required are a monitor that that will display graphs and a printer that will make hard copies of them.

The diskette in this book contains 1-2-3 worksheets. You will need to copy this diskette (termed the original diskette) to one or more working diskettes or to a hard disk. There are two reasons for doing this. First, it is desirable to save the original diskette so that it will not wear out through repeated use, and, second, you might inadvertently erase or alter a worksheet in a way that you do not intend, so you need to be able to retrieve an unaltered worksheet from an original diskette. (I generally keep two or three working diskettes.) Because the original diskette is write protected, it is difficult to damage while you are copying, but, if you are uncertain about the procedure, you may want to consult your computer manual and practice on other diskettes.

To copy the original diskette to a working diskette, follow the instructions with your computer. For the IBM Personal Computer with two diskette drives, first format a diskette. The command to copy and verify 1-2-3 worksheet files is

COPY *.* B:/V.

Alternatively, you can copy the original diskette to a directory on a hard disk.

The worksheets take up about 310,000 of the some 360,000 characters that can be stored on a floppy diskette.

CONVENTIONS

Long worksheets are programmed for manual recalculation. To recalculate, press the function key (F9).

All references to particular keys are to the keyboard of the IBM Personal Computer. Cell references are in capital letters and numbers, for example, D7. For ease in reference, I have numbered the rows of each worksheet in column A and have left column B empty. Columns are labeled wherever necessary.

Most formulae follow the notation in Peters's (1987) book, except for compound interest factors, for which I have adopted Stermole and Stermole's (1987) notation. In general, metric units and U.S. dollars are used.

For clarity, I have labeled most of the variables used in the worksheets, using 1-2-3's /**R**ange **N**ame **L**abel command. In the individual worksheets, the variables are listed in 1-2-3's alphabetical order, as provided by the command /**R**ange **N**ame **T**able.

Lotus 1-2-3 limits labels to 15 upper-case alphabetic and certain other characters; within this limitation my intent is to make the labels easy to read. Toward this end, I have used these conventions:

*Words, letters, and numbers within a label are separated by periods (.) rather than by symbols like slashes (/) and dashes (-) that can be confused with mathematical operators.

*Compound interest factors are preceded by a poundsign (#).

*Labels for dollar values are preceded by a exclamation mark (!), rather than a dollar sign, because 1-2-3 uses a dollar sign for a particular purposc (to set cell references).

*Percentage factors are preceded by a percent sign (%).

*Labels for metals start with the chemical symbol, for example AG.

Macros, which are programs written in 1-2-3's command language, are used in Chapters 6 and 7. They are listed in the worksheets and named in alphabetical order.

CONCLUSIONS

As you practice with the worksheets in this book, you will doubtless learn many ways to adopt them for your particular purposes. They will also suggest other worksheets that you devise. 1-2-3 worksheets and templates are being published in increasing numbers in geology and mining engineering.

What precautions need you take? For at least two reasons, it is difficult to locate mistakes in worksheets: (1) worksheets do not separate formulae and data and (2) macros are, at present, difficult to read and correct. This situation, we can hope is temporary. Moreover, one always need to be aware of the choices possible to solve a particular problem, one or another spreadsheet, perhaps a statistical-analysis package, or a high-level language.

In the future, I anticipate that it will be easier to interface the various types of computational aids with one another than at present. And more functions will be developed for 1-2-3 that will make scientific, statistical, and engineering computing more routine that at present.

CHAPTER 2

Confidence Intervals

In this chapter, we will use 1-2-3 to calculate confidence intervals for a population mean. The principles and methods are explained by Koch and Link (1970, p. 79-104), Peters (1987), p. 476), and by standard statistical texts.

2.1 CONFIDENCE INTERVALS FOR A FICTITIOUS DATA SET

Suppose that you have drilled five exploration diamond-drillholes in the Middle Tennessee zinc district. Suppose further that the zinc assays are 2, 4, 6, 6, and 7%, as recorded in Part 1 of worksheet C2A.WK1.

In Part 2, we calculate a confidence limit at a 10-percent risk level for the data in Part 1. Line 26 records the number of drillholes; line 28 the arithmetic average (sample mean); line 30 the sample variance; and line 32 the sample standard deviation. The value for Student's t is entered in line 34 from a standard table present in statistical textbooks (also available in Koch and Link, 1970, p. 346). Line 36 is D, the distance that is one-half the width of the confidence interval; line 38 is UCL, the upper confidence limit; and line 40 is LCL, the lower confidence limit. For this data set, the sample variance is exactly 4, and the sample standard deviation is exactly 2. Therefore, it is easy to check the 1-2-3 arithmetic that calculates these statistics. Because 1-2-3's function @VAR calculates a population variance rather than a sample variance, the formula in line 30 is corrected to obtain the required sample variance. Part 3 of worksheet C2A.WK1 is the table of labels used to make the formulae easy to read.

2.2 CONFIDENCE INTERVALS FOR DATA FROM THE ELMWOOD, TENNESSEE ZINC DEPOSIT

The data in the previous section are fictitious, selected for their simplicity. In Part 4, we repeat the calculations for data from six surface boreholes drilled to evaluate the Elmwood, Tennessee zinc deposit. These holes are six of 121 holes drilled by the New Jersey Zinc company to evaluate the deposit (Callahan, 1977); the data are from Chapter 6 (worksheet C6B.WK1), where the geological situation is discussed. The confidence interval, from 4.10 to 5.60 percent zinc, includes the mined grade of 4.2 percent for the first year of production (Koch and Schuenemeyer, 1982, p. 660).

2.3 CONFIDENCE INTERVALS FOR THE DON TOMAS VEIN, FRISCO MINE

The Don Tomas vein of the Frisco mine, Chihuahua, Mexico contained a large and rich ore shoot some 600 m long and 40 m high (Koch and Link, 1964); Figure 2.1 is a vertical longitudinal section. The vein was discovered and explored by diamond drilling.

Part 1 of worksheet C2B.WK1 tabulates data for 18 diamond-drillholes bored through the vein. (The units, meter-grams per tonne, and meter-percent per tonne are appropriate for evaluating narrow tabular ore bodies.) Part 2 gives the calculations and confidence intervals for the five metals that were produced from this vein. As explained by Koch and Link (1970, p. 100), the confidence limits include the grade of ore mined

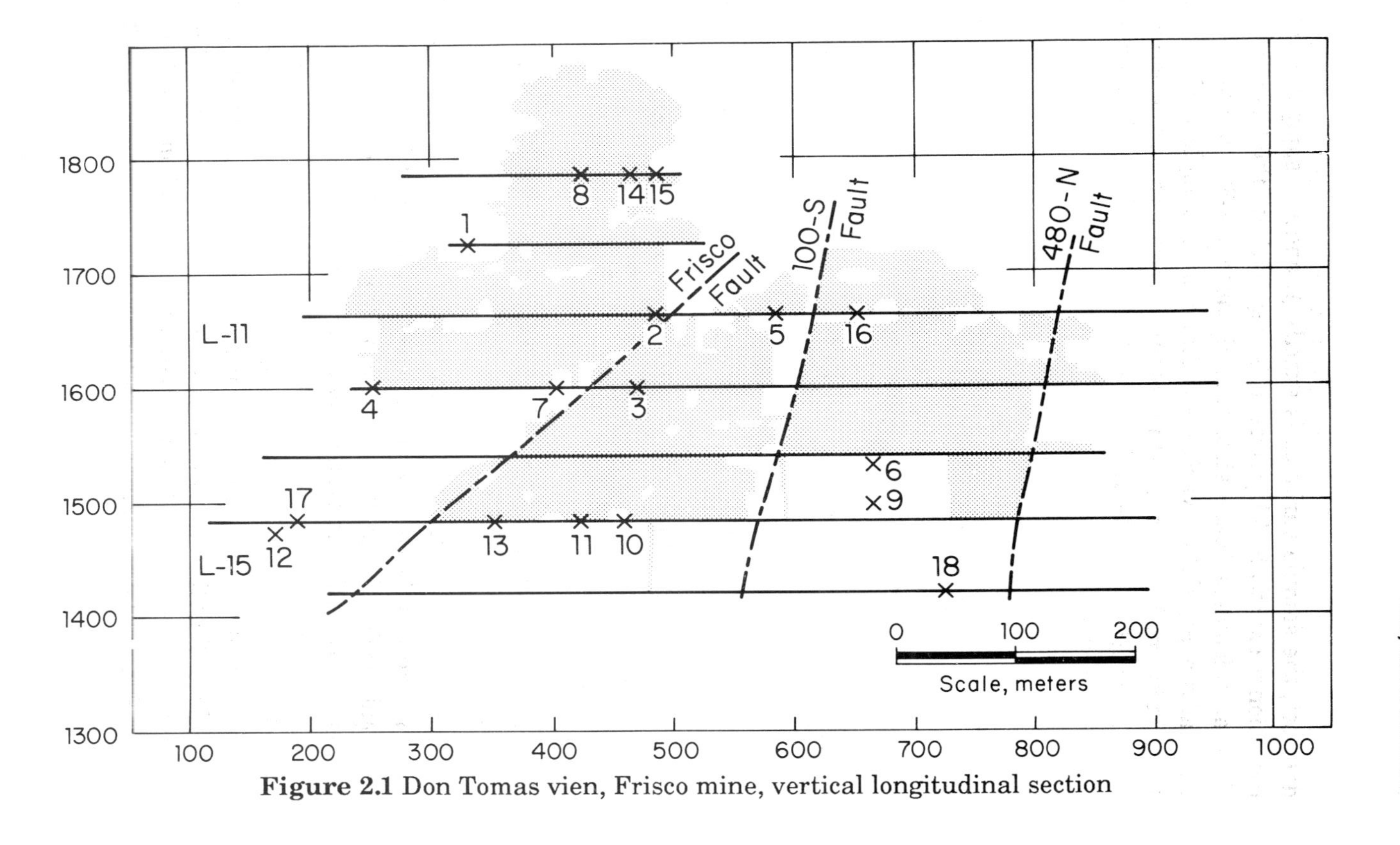

Figure 2.1 Don Tomas vien, Frisco mine, vertical longitudinal section

as established from 1,829 drift samples taken after the vein was developed for mining.

Usually, confidence intervals narrow as more data (observations, in statistical nomenclature) are obtained, corresponding in this example to more holes being drilled. They narrow for two reasons: (1) in the equation used to calculate D in line 51, the number of observations, N, in the denominator, increases, and (2) the value of t, in the numerator, decreases. Figure 2.2 graphs this change for lead. The interval width narrows from 2 to 13 holes as expected. It then widens at hole 14, because the grade of lead in this hole is so high, an illustration of statistical fluctuation. The interval width for holes 15 to 18 continues to narrow progressively.

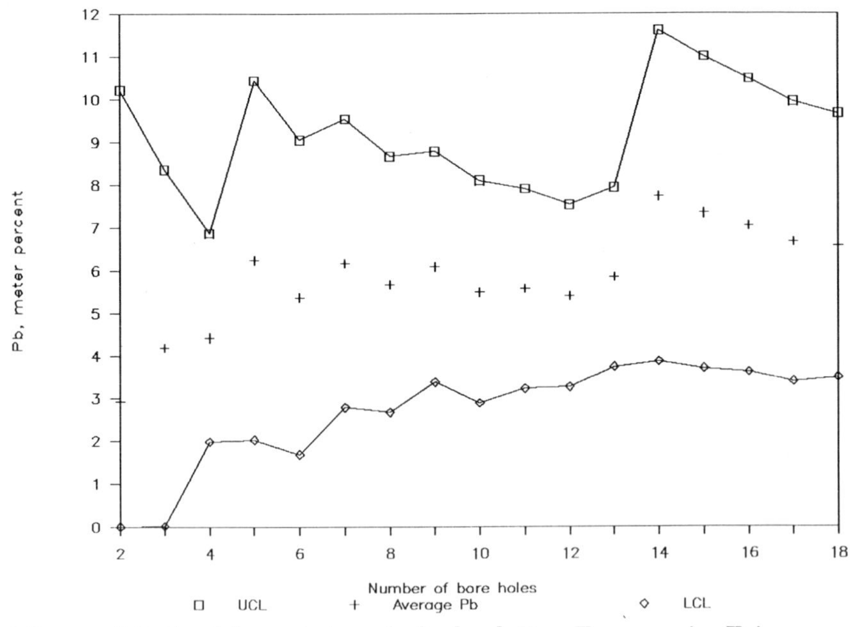

Figure 2.2 Confidence intervals for lead, Don Tomas vein, Frisco mine.

Worksheet C2A Confidence intervals for Elmwood mine

```
1   1 Dec 87, file C2A.WK1 on disks GSK 059-061
2
3  Worksheet C2A.WK1. Confidence intervals for the Elmwood
mine
4
5   1. Fictitious assay data from drilling in Middle
Tennessee
6
7  ---------
8      Zn,
9       %
10 ---------
11         2
12         4
13         6
14         6
15         7
16 ---------
17
18
19 2. Calculation of a confidence interval for the popula-
tion mean
20
21      C        D
22 ------------------
23  Item        Zn,
24               %
25 ------------------
26  N              5
27
28  AVG            5
29
30  VAR            4
31
32  STD            2
33
34  T(5%)      2.132
35
36  D           1.91
```

Worksheet C2A *(continued)*

```
37
38  UCL              6.91
39
40  LCL              3.09
41
42  ------------------
43
44
45  3.  Table of labels
46
47  AVERAGE               D28
48  D                     D36
49  HOLES                 C11..C15
50  N                     D26
51  STD.DEVIATION         D32
52  T                     D34
53  VARIANCE              D30
54
55
56  4.  Elmwood mine data (from worksheet C6B.WK1).
57
58  ---------
59     Zn,
60      %
61  ---------
62       3.6
63       6.0
64       5.2
65       5.9
66       5.5
67       3.4
68       3.6
69       5.6
70   ---------
71
72
```

Worksheet C2A *(continued)*

```
73  2.  Calculation of a confidence interval for the
population mean
74
75      C        D
76 ------------------
77  Item         Zn,
78               %
79 ------------------
80  N               8
81
82  AVG          4.85
83
84  VAR          1.25
85
86  STD          1.12
87
88  T(5%)       1.895
89
90  D            0.75
91
92  UCL            5.60
93
94  LCL          4.10
95
96 ------------------
```

Worksheet C2B Confidence intervals for Frisco mine

1 11 Dec 87, file C2B.WK1 on disks GSK 059-061
2
3 Worksheet C2B.WK1. Confidence intervals for the Frisco mine
4
5 1. Assay data from diamond-drillholes bored through the
6 Don Tomas vein (from Koch and Link, 1964, p.24).
7
8

9	C	D	E	F	G	H	I	J
11	Hole	Au,	Ag,	Pb,	Cu,	Zn,	x,	y,
12	No.	m-g/T	m-g/T	m-%	m-%	m-%	metrs	metrs
14	1	0.15	16.8	1.77	0.090	1.44	328	1,728
15	2	0.24	115.2	4.08	0.072	13.20	496	1,665
16	3	1.20	384.0	6.72	0.120	5.88	460	1,603
17	4	0.00	172.8	5.16	0.240	4.92	250	1,603
18	5	0.00	893.2	13.44	0.700	5.11	583	1,665
19	6	0.00	80.0	1.04	0.080	1.04	665	1,518
20	7	1.30	239.2	10.92	0.416	5.98	400	1,603
21	8	0.07	37.8	2.24	0.238	4.76	422	1,787
22	9	0.00	708.0	9.36	0.360	8.64	665	1,499
23	10	0.00	4.0	0.24	0.060	0.24	458	1,483
24	11	0.00	88.0	6.25	0.190	12.50	413	1,483
25	12	1.08	252.0	3.60	0.594	9.36	166	1,473
26	13	0.00	261.0	11.10	0.250	25.40	346	1,483
27	14	0.00	294.8	32.34	1.210	36.08	465	1,787
28	15	0.00	45.0	1.98	0.315	4.59	483	1,787
29	16	0.49	256.2	2.52	0.140	3.22	652	1,665
30	17	0.00	170.1	0.63	0.903	10.71	183	1,483
31	18	0.35	987.0	4.90	0.455	1.75	727	1,424

33 Note: m-g/T = meter-grams per tonne; m-% = meter-percent per tonne

Worksheet C2B *(continued)*

```
34
35
36  2.  Calculation of a confidence interval for the
population mean
37   ----------------------------------------------
38  Item     Au,       Ag,      Pb,      Cu,       Zn,
39         m-g/T    m-g/T    m-%      m-%       m-%
40 ---------------------------------------------
41  N        18        18      18       18        18
42
43  AVG     0.27       278   6.57     0.36      8.60
44
45  VAR     0.20   85,901  56.68     0.10     82.58
46
47  STD     0.45       293   7.53     0.32      9.09
48
49  T(5%)   1.74         2   1.74     1.74      1.74
50
51  D       0.18       120   3.09     0.13      3.73
52
53  UCL     0.46       398   9.66     0.49     12.33
54
55  LCL     0.09       158   3.48     0.23      4.87
56
57------------------------------------------------
58
59
60  3.  Table of labels
61
62  AVERAGE               D43
63  D                     D51
64  HOLES                 D14..D31
65  N                     D41
66  STD.DEVIATION         D47
67  T                     D49
68  VARIANCE                D45
```

CHAPTER 3

Fitting a Straightline To Exploration Data

In this chapter, we will fit a straightline to exploration data through the statistical method of linear regression, which is explained by Koch and Link (1971, p. 7-15) and by standard statistical textbooks. This method determines the equation of the straightline with the property that the sum of the squared vertical deviations of the points from the line is a minimum.

Linear regression is also useful for problems of ore-deposit evaluation, for instance to compare different methods of sampling or assaying.

3.1 GOLD PLACER DATA FROM MANHATTAN, NEVADA

Worksheet C3A.WK1 contains data for gold nuggets collected in a stream near Manhattan, Nevada (Ferguson, 1916). The following hypothesis will be tested: that gold percentage rises with increasing distance from the place where the stream crosses a lode gold deposit that is the source of the nuggets. The supposition is that, as the nuggets travel downstream, silver and base metals are leached from them.

Columns C and D of the data table give the paired original values for distance and gold percentage. After the data table, I did a linear regression using the /Data Regression command. The output, which is presented in lines 49 to 57, is explained in the references cited in the previous paragraph and in 1-2-3 manuals.

In lines 63 to 66, we set up a table of labels. CONSTANT, the y-intercept, is in cell F50, and X.COEFFICIENT, the slope of the line, is in cell E56. Remembering the equation for a straightline,

$$\text{y-hat} = ax + b,$$

where y-hat is the estimated gold percentage, x is the distance, and a and b are constants named the slope and intercept, respectively, we obtain for cell E12 the formula,

($X.COEFFICIENT)*(C12)+($CONSTANT).

By copying this formula into cells E13..E43, we obtain the fitted y-hat values corresponding to the original data points. (The y-hat values are the gold values predicted by the linear equation for each of the x-values of distance.)

Figure 3.1 shows the original data points and the plotted straight-line.

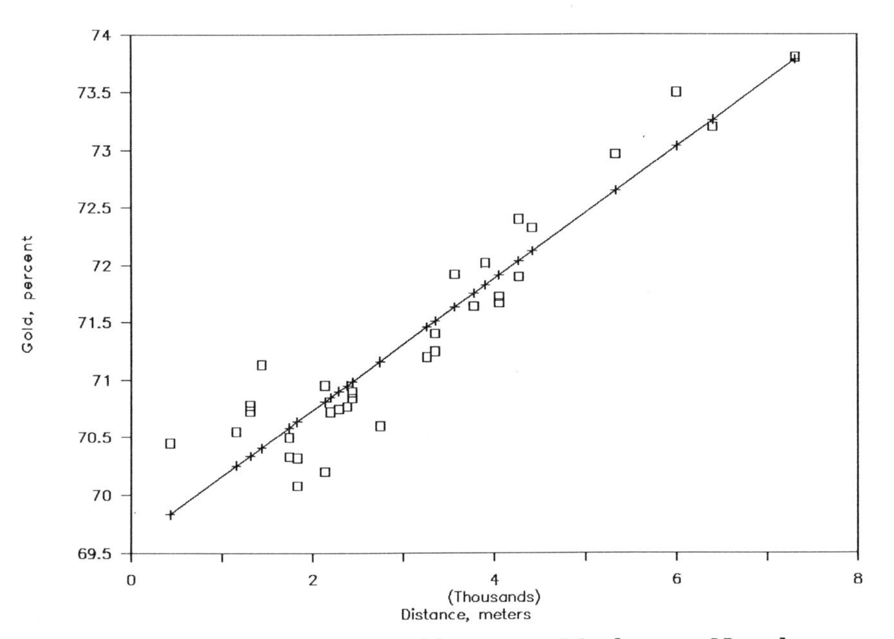

Figure 3.1 Linear regression, gold nuggets, Manhattan, Nevada.

Worksheet C3A Linear regression

```
 1  15 Sep 87, file C3A.WK1 on disks GSK 056-0
 2
 3 Worksheet C3A.WK1.  Linear regression
 4
 5  Part 1.  Table of data
 6
 7            C              D          E          F
 8 ______________________________
 9           x,              y,      y-hat,
10           m               %          %
11 ______________________________
12              427     70.45      69.83
13             1158     70.55      70.25
14             1311     70.73      70.34
15             1311     70.78      70.34
16             1433     71.13      70.41
17             1737     70.33      70.58
18             1737     70.50      70.58
19             1829     70.08      70.64
20             1829     70.32      70.64
21             2134     70.95      70.81
22             2134     70.20      70.81
23             2195     70.72      70.85
24             2286     70.75      70.90
25             2377     70.77      70.95
26             2438     70.84      70.99
27             2438     70.90      70.99
28             2743     70.60      71.16
29             3261     71.20      71.46
30             3353     71.40      71.51
31             3353     71.25      71.51
32             3566     71.92      71.63
33             3780     71.64      71.76
34             3901     72.02      71.83
35             4054     71.73      71.91
36             4054     71.67      71.91
37             4267     72.40      72.04
38             4267     71.90      72.04
```

Worksheet C3A *(continued)*

```
39           4420     72.33    72.12
40           5334     72.96    72.65
41           6005     73.50    73.03
42            6401    73.20     73.26
43            7315    73.80     73.78
44 ---------------------------
45
46
47 Part 2. 1-2-3's linear regression
48 ---------------------------
49                  Regression Output:
50  Constant                          69.58880
51  Std Err of Y Est                  0.342396
52  R Squared                         0.884272
53  No. of Observations                     32
54  Degrees of Freedom                      30
55
56  X Coefficient(s)        0.000573
57  Std Err of Coef.        0.000037
58
59
60 ---------------------------
61
62
63  3.  Table of labels
64
65  CONSTANT        F50
66   X.COEFFICIENT   E56
```

CHAPTER 4

Analysis Of Variance

If we have several sets of exploration data, we may want to compare them to learn whether they come from bodies with similar mineralization, as measured in quantitative terms that may be grades of ore, geochemical values, tonnages, or one of many other variables. In statistical nomenclature, we are comparing sample data to determine whether populations are evidently the same or different. For example, we may have sampled several Nevada prospects for gold. Do the data support the hypothesis that the gold content of the orebodies is the same, recognizing that the sampling provides only an estimate of the true grades?

We can answer this question by making a one-way analysis of variance, as explained by Koch and Link (1970, p. 132-201) and by standard statistical textbooks.

The one-way analysis of variance is only the simplest form of this powerful method of statistical analysis. Using 1-2-3, other forms of the analysis of variance can be programmed. In this book, I provide a worksheet for one other form, the randomized block analysis of variance. This analysis is particularly useful in mining geology for comparing various methods of sampling, sample preparation, or chemical analysis. Li (1964) provides many other easy-to-program models.

Sections 4.1 and 4.2 explain the one-way analysis of variance; sections 4.3 and 4.4 explain the randomized-block analysis of variance.

4.1 ONE-WAY ANALYSIS OF VARIANCE

Worksheet C4A.WK1 for the one-way analysis of variance follows the form of Li (1964, p. 194-196).

Part 1 is a table of data; there are three groups, numbered 1 to 3, and eight values, of which two are in the first sample, two in the second, and four in the third. Pretend that we drilled holes in three areas in the Middle Tennessee zinc district and obtained these zinc analyses in the eight holes.

Part 2 is the one-way analysis of variance. 1-2-3 calculates all of the entries, except for the tabled value 3.78 in cell I44, which is the F-value at the 10% level with 2 and 5 degrees of freedom from Koch and Link (1970, p. 348) or from any statistical textbook. Because the calculated F-value is larger than the tabled F-value, we conclude that the three groups of samples represent groups with significantly different average grades. In terms of our example, this small amount of drilling suggests that the three areas in Middle Tennessee have significantly different grades of zinc mineralization.

Part 3 is the table of labels.

Part 4 details the calculations. Part 4.A lists the sums, counts, and means; Part 4.B is the table to compute the elements, I, II, and III, which are used to calculate the analysis of variance. Part 4.C gives the squared matrix of the table of data. The table provides for as many as six data groups (samples, in statistical terminology) and as many as 20 data points (observations in statistical terminology) per group. If the data table has too few rows, you can increase the number with the command /Worksheet Insert Row, in order to avoid changing the formulae. Increasing the number of columns will require minor modifications of the formulae.

4.2 COMPARING VEIN SETS IN THE FRISCO MINE, CHIHUAHUA, MEXICO

In the Frisco mine, Chihuahua, Mexico, veins fall into one of four sets according to their strike and dip (Koch, 1956, p. 14-15). Part 1 of worksheet C4B.WK1 gives the percentage of the vein that is ore for the four sets. Part 2 gives the one-way analysis of variance. Because the calculated F-value of 3.2 is larger than the tabled F-value of 2.31, we conclude that evidently the percentage of ore in the vein sets differs significantly. As discussed in my original paper, this difference presumably stems from the tectonic history of the ore deposit.

4.3 RANDOMIZED-BLOCK ANALYSIS OF VARIANCE

Worksheet C4C.WK1 for the randomized-block analysis of variance follows the form of Li (1964, p. 244).

Part 1 is a table of data, consisting of readings of pH from the top, middle, and bottom of six core samples of soil. There are three groups, numbered 1 to 3 (corresponding to top, middle, and bottom), and six values in each group (corresponding to the six samples). The left-hand row contains the sample totals T(r), used to calculate element II in cell F86.

The purpose of this analysis of variance is to test whether the pH in the top, middle, and bottom soil samples is the same; in statistical terms, we test the hypothesis H(0) that there is no treatment effect against the alternative hypothesis H(1) that there is a treatment effect.

Part 2 is the randomized-block analysis. As for the one-way analysis of variance, 1-2-3 calculates all of the entries, except for the tabled values of F in cells I44 and I45. Because the calculated F-value, 31.79, is larger than the tabled F-value, 2.92, we conclude that the pH in the top, middle, and bottom soil samples are significantly different, or, in statistical terms, that there is a treatment effect.

This analysis of variance also can be used to investigate whether the six places sampled, on the average, have significantly different soil pH values. In statistical terms, the hypothesis H(0) that the soil is the same at the six places, that is, no replication effect exists, is compared with the alternative hypothesis H(1) that the soil is significantly different. Because the calculated F-value to test this hypothesis is 6.43, which is larger than the tabled F-value of 2.52, we conclude that the soil is significantly different.

The labels in Parts 1 and 2 are general ones. To present this example analysis for publication, I would change the group numbers in line 12 to indicate the soil type, top, middle, and bottom. And, in cell C44, the appropriate label would be "Among soil depths," and in cell C45, "Among core samples." This relabeling, easily done in 1-2-3, will help you and your users understand the geological significance of an analysis of variance.

Parts 3 and 4 correspond to those of worksheet C4A.WK1 for the one-way analysis of variance, with a few changes in detail.

4.4 COMPARING METHODS OF SAMPLE PREPARATION FOR GOLD ORE

Given geological samples of gold ore to be assayed, what methods of sample preparation yield reliable results at the lowest cost? That is, how best can time and money be allocated among crushing, grinding, pulverizing, splitting, and assaying? Statistical methods to answer this general question include the randomized-block analysis of variance. In this section, I use data from the Homestake Mine, Lead, South Dakota (Koch and Link, 1972, p.13).

Worksheet C4D.WK1 analyzes data from three methods of sample preparation: crushing, grinding, and pulverizing, which are listed as groups 1, 2, and 3, respectively in Part 1 of the worksheet.

Part 2 gives the analysis of variance. Because the F-value of 0.58 for replication variability is smaller than the tabled value of 2.92, we conclude that there is no significant difference among the three methods of sample preparation for these data.

The randomized-block analysis of variance is only one of several statistical methods that we used in the original paper.

Worksheet C4A One-way analysis of variance

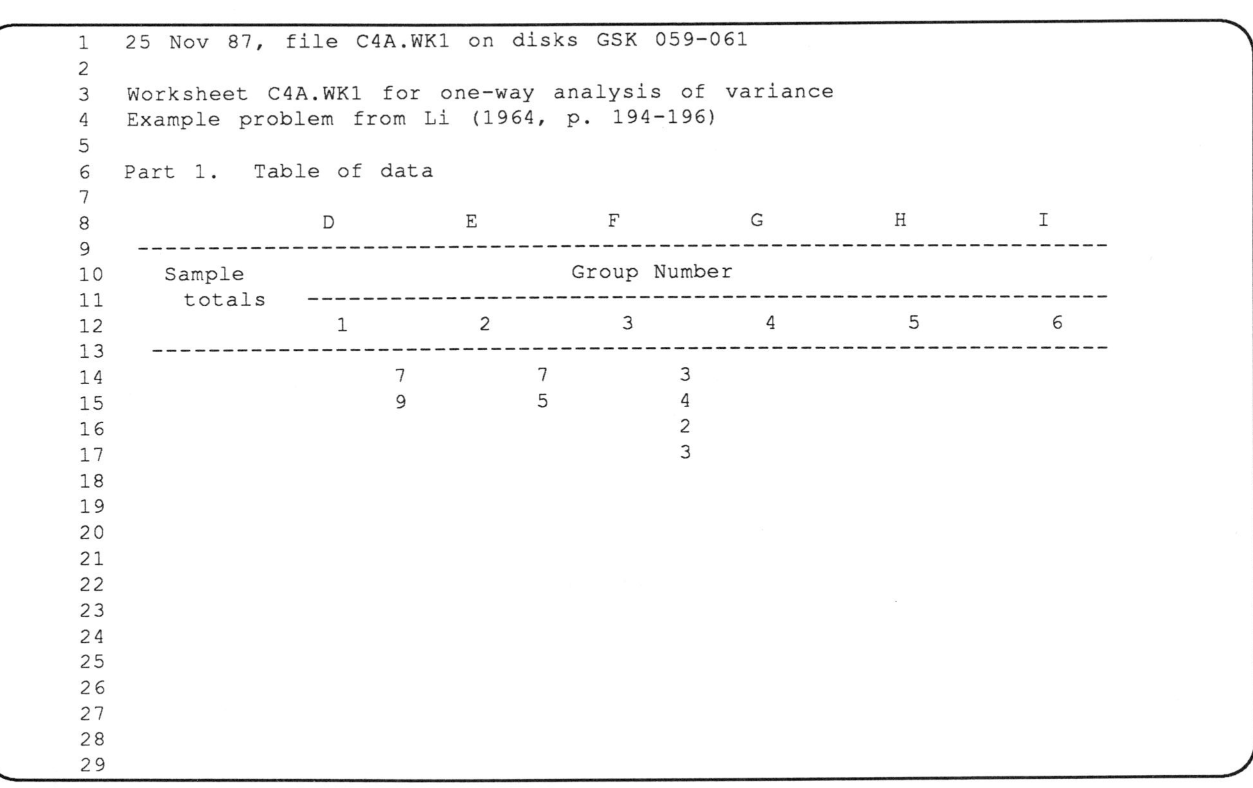

```
1   25 Nov 87, file C4A.WK1 on disks GSK 059-061
2
3   Worksheet C4A.WK1 for one-way analysis of variance
4   Example problem from Li (1964, p. 194-196)
5
6   Part 1.  Table of data
7
8                  D          E          F          G          H          I
9   ---------------------------------------------------------------------------
10     Sample                          Group Number
11      totals  ---------------------------------------------------------------
12                  1          2          3          4          5          6
13   ---------------------------------------------------------------------------
14                      7          7          3
15                      9          5          4
16                                            2
17                                            3
18
19
20
21
22
23
24
25
26
27
28
29
```

Worksheet C4A *(continued)*

```
30
31
32
33
34   --------------------------------------------------------------------------
35
36
37   Part 2.  Analysis of variance
38
39                              E          F           G          H          I
40   --------------------------------------------------------------------------
41                         Sum of  Degrees of    Mean
42   Source of variation  squares  freedom      square         F        F(10%)
43   --------------------------------------------------------------------------
44   Among-samples             36           2       18.0       15.0       3.78
45   Within-samples             6           5        1.2
46   Total                     42           7
47   --------------------------------------------------------------------------
48
49
50   Part 3. Table of labels
51
52   DF.AMONG             F44
53   DF.TOTAL             F46
54   DF.WITHIN            F45
55   G                    H35
56   G*G                  F88
57   I                    F88
58   II                   F89
```

Worksheet C4A *(continued)*

```
59  III                    F90
60  K                      D79
61  MS.AMONG               G44
62  MS.WITHIN              G45
63  N                      E88
64  SS.AMONG               E44
65  SS.TOTAL               E46
66  SS.WITHIN              E45
67
68
69
70
71   Part 4.   Computation
72
73   4.A.   Sums, counts, and means
74
75  SUM(W)              16          12          12
76  COUNT(W)             2           2           4
77  AVG(W)               8           6           3
78  SUM(W)^2/N         128          72          36
79  K                    3
80
```

Worksheet C4A *(continued)*

```
81  4.B.  Computing table for one-way analysis of variance
82  (from Koch and Link, 1970, p. 141)
83
84   ---------------------------------------------------
85    Type       Total of Number of  Squares/  Element
86    of sum    squares   observ.   observ.
87   ---------------------------------------------------
88  Grand           1600          8        200        (I)
89  Sample                                 236       (II)
90  Observ.          242          1        242      (III)
91   ---------------------------------------------------
92
```

Worksheet C4A *(continued)*

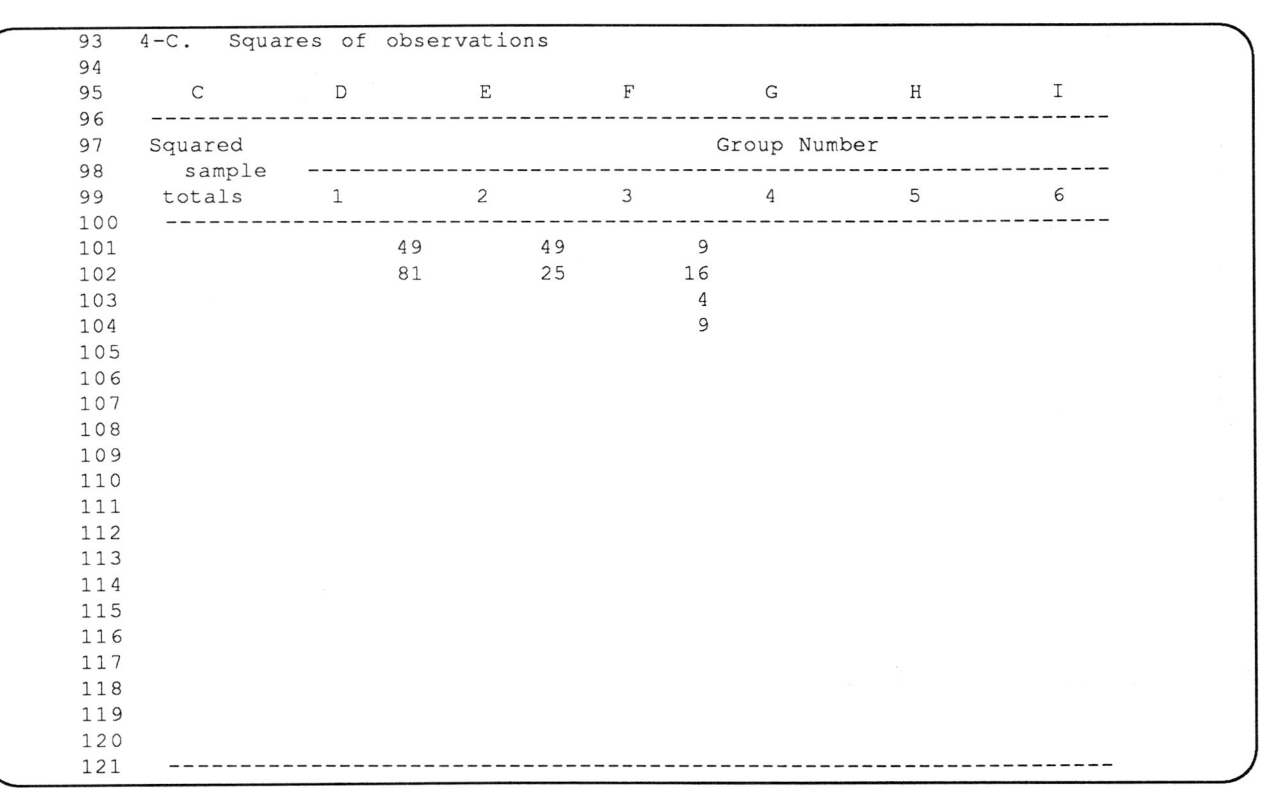

```
 93  4-C.   Squares of observations
 94
 95     C          D          E          F          G          H          I
 96   ---------------------------------------------------------------------
 97   Squared                                      Group Number
 98     sample   -----------------------------------------------------------
 99    totals     1          2          3          4          5          6
100    --------------------------------------------------------------------
101                   49         49          9
102                   81         25         16
103                                          4
104                                          9
105
106
107
108
109
110
111
112
113
114
115
116
117
118
119
120
121    --------------------------------------------------------------------
```

Worksheet C4B One-way analysis of variance, Frisco mine

```
 1   25 Nov 87, file C4B.WK1 on disks GSK 059-061
 2
 3   Worksheet C4B.WK1. One-way analysis of variance, Frisco mine
 4   Comparison of four vein sets in the Frisco mine
 5
 6   Part 1.  Table of data
 7
 8                  D          E          F          G          H          I
 9    -------------------------------------------------------------------------
10     Sample                        Group Number
11      totals  ---------------------------------------------------------------
12                  1          2          3          4          5          6
13    -------------------------------------------------------------------------
14                     83         10         53         76
15                     61         43         58         50
16                     56         26         35         60
17                     57         15         19         50
18                     25         40                    64
19                     67         67                    33
20                     67                               67
21                     33                               80
22                     20                               83
23                                                      90
24                                                      40
25
26
27
28
29
```

Worksheet C4B *(continued)*

```
30
31
32
33
34   ---------------------------------------------------------------------------
35
36
37   Part 2.   Analysis of variance
38
39                             E          F          G          H          I
40   ---------------------------------------------------------------------------
41                         Sum of  Degrees of    Mean
42   Source of variation  squares   freedom     square         F        F(10%)
43   ---------------------------------------------------------------------------
44   Among-samples           3,813          3    1270.9        3.2        3.78
45   Within-samples         10,185         26     391.7
46   Total                  13,998         29
47   ---------------------------------------------------------------------------
48
49
50   Part 3. Table of labels
51
52   DF.AMONG             F44
53   DF.TOTAL             F46
54   DF.WITHIN            F45
55   G                    H35
56   G*G                  F88
57   I                    F88
58   II                   F89
```

Worksheet C4B *(continued)*

```
59   III                    F90
60   K                      D79
61   MS.AMONG               G44
62   MS.WITHIN              G45
63   N                      E88
64   SS.AMONG               E44
65   SS.TOTAL               E46
66   SS.WITHIN              E45
67
68
69
70
71    Part 4.   Computation
72
73    4.A.   Sums, counts, and means
74
75   SUM(W)           469        201        165        693
76   COUNT(W)           9          6          4         11
77   AVG(W)     52.111111       33.5      41.25         63
78    SUM(W)^2/N24440.111     6733.5    6806.25      43659
79   K                  4
80
```

Worksheet C4B *(continued)*

```
81  4.B.  Computing table for one-way analysis of variance
82  (from Koch and Link, 1970, p. 141)
83
84   ----------------------------------------------------
85    Type       Total of Number of  Squares/  Element
86    of sum     squares    observ.   observ.
87   ----------------------------------------------------
88  Grand         2334784         30 77826.133         (I)
89  Sample                            81638.861        (II)
90  Observ.         91824          1     91824       (III)
91   ----------------------------------------------------
92
```

Worksheet C4B *(continued)*

```
 93  4-C.   Squares of observations
 94
 95     C          D          E          F          G          H          I
 96   ------------------------------------------------------------------------
 97   Squared                                      Group Number
 98     sample    ------------------------------------------------------------
 99    totals      1          2          3          4          5          6
100   ------------------------------------------------------------------------
101                   6889        100       2809       5776
102                   3721       1849       3364       2500
103                   3136        676       1225       3600
104                   3249        225        361       2500
105                    625       1600                  4096
106                   4489       4489                  1089
107                   4489                             4489
108                   1089                             6400
109                    400                             6889
110                                                    8100
111                                                    1600
112
113
114
115
116
117
118
119
120
121   ------------------------------------------------------------------------
```

Worksheet C4C Randomized-block analysis of variance

```
 1   1 Dec 87, file C4C.WK1 on disks GSK 059-061
 2
 3   Worksheet C4C.WK1 for randomized-block analysis of variance
 4   Example problem from Li (1964, p. 244)
 5
 6   Part 1.  Table of data
 7
 8       C          D          E          F          G          H          I
 9    ------------------------------------------------------------------------
10      Sample                           Group Number
11       totals  -------------------------------------------------------------
12                  1          2          3          4          5          6
13    ------------------------------------------------------------------------
14        22.3       7.5        7.6        7.2
15        21.0       7.2        7.1        6.7
16        21.5       7.3        7.2        7.0
17        21.9       7.5        7.4        7.0
18        22.4       7.7        7.7        7.0
19        22.2       7.6        7.7        6.9
20
21
22
23
24
25
26
27
28
29
```

Worksheet C4C *(continued)*

```
30
31
32
33
34   ----------------------------------------------------------------------
35
36
37   Part 2.   Analysis of variance
38
39                             E          F          G          H          I
40   ----------------------------------------------------------------------
41                        Sum of  Degrees of    Mean
42   Source of variation  squares  freedom      square         F     F(10%)
43   ----------------------------------------------------------------------
44   Replication            0.4894          5   0.09789      6.43       2.52
45   Treatment              0.9678          2   0.48389     31.79       2.92
46   Residual               0.1522         10   0.01522
47   Total                  1.6094        17
48   ----------------------------------------------------------------------
49
50   Part 3. Table of labels
51
52   DF.REPLICATION        F44
53   DF.RESIDUAL           F46
54   DF.TOTAL              F47
55   DF.TREATMENT          F45
56   G*G                   D85
57   I                     F85
58   II                    F86
```

Worksheet C4C *(continued)*

```
59  III                    F87
60  IV                     F88
61  K                      E86
62  MS.REPLICATION         G44
63  MS.RESIDUAL            G46
64  MS.TREATMENT           G45
65  N                      D76
66  SS.REPLICATION         E44
67  SS.RESIDUAL            E46
68  SS.TOTAL               E47
69  SS.TREATMENT           E45
70
71
72  4.A.   Sums and counts
73
74  SUM(W)           44.8       44.7       41.8
75  SUM(W)^2      2007.04    1998.09    1747.24
76  COUNT(W)            6          6          6
77
78  4.B.   Computing table for randomized block
79  (from Koch and Link, 1970, p. 159)
80
```

Worksheet C4C *(continued)*

```
81   ------------------------------------------------
82     Type       Total of Number of  Squares/  Element
83     of sum     squares   observ.    observ.
84   ------------------------------------------------
85   Grand       17239.69         18  957.7606       (I)
86   Replicat.    2874.75          3  958.2500      (II)
87   Treat.       5752.37          6  958.7283     (III)
88   Observ.       959.37          1  959.3700      (IV)
89   ------------------------------------------------
90
91
92
```

Worksheet C4C *(continued)*

```
 93   4-C.   Squares of observations
 94
 95      C          D          E          F          G          H          I
 96   ----------------------------------------------------------------------
 97   Squared                                    Group Number
 98     sample   -----------------------------------------------------------
 99    totals        1          2          3          4          5          6
100    -------------------------------------------------------------------
101      497.29      56.25      57.76      51.84
102      441.00      51.84      50.41      44.89
103      462.25      53.29      51.84      49.00
104      479.61      56.25      54.76      49.00
105      501.76      59.29      59.29      49.00
106      492.84      57.76      59.29      47.61
107
108
109
110
111
112
113
114
115
116
117
118
119
120
121    -------------------------------------------------------------------
```

Worksheet C4D Randomized-block analysis of variance, Homestake mine

```
 1  1 Dec 87, file C4D.WK1 on disks GSK 059-061
 2
 3  Worksheet C4D.WK1. Randomized-block analysis of variance for
 4     Homestake mine data.  Data from Koch and Link (1972, p. 13).
 5
 6  Part 1.  Table of data
 7
 8      C          D          E          F          G          H          I
 9   ----------------------------------------------------------------------
10     Sample                         Group Number
11      totals  ----------------------------------------------------------
12                 1          2          3          4          5          6
13   ----------------------------------------------------------------------
14     23.797      5.198      6.254    12.3445
15      5.405     0.3255     2.1327     2.9472
16      3.965     0.0872      0.001     3.8765
17      0.367     0.2345      0.001     0.1315
18      3.698      0.001      0.001     3.6955
19     39.803     23.956     8.8585     6.9887
20      3.549     2.4835     0.3767      0.689
21      0.969      0.429      0.334     0.2057
22      0.879      0.001     0.2147     0.6633
23     11.449     8.3112      0.894      2.244
24      2.236     0.1947      0.385     1.6565
25     18.930      4.299     6.4242      8.207
26      4.751      0.645     1.3372      2.769
27     45.712     1.3077    17.1602     27.244
28     37.878    22.9445      7.779     7.1545
```

Worksheet C4D *(continued)*

```
29      25.885    18.9772       3.184       3.724
30      58.338    11.3167     25.5457     21.4755
31       2.447      0.832       0.734       0.881
32       0.003      0.001       0.001       0.001
33      39.373    22.4892      7.0732      9.8102
34   ---------------------------------------------------------------------
35
36
37  Part 2.   Analysis of variance
38
39                           E          F          G          H          I
40   ---------------------------------------------------------------------
41                          Sum of  Degrees of   Mean
42  Source of variation   squares   freedom     square        F        F(10%)
43   ---------------------------------------------------------------------
44  Replication          2156.0567        19 113.47667      3.78        2.52
45  Treatment              34.7959         2  17.39795      0.58        2.92
46  Residual             1141.2244        38  30.03222
47  Total                3332.0770        59
48   ---------------------------------------------------------------------
49
50  Part 3. Table of labels
51
52  DF.REPLICATION        F44
53  DF.RESIDUAL           F46
54  DF.TOTAL              F47
55  DF.TREATMENT          F45
56  G*G                   D85
57  I                     F85
```

Worksheet C4D *(continued)*

```
58  II                   F86
59  III                  F87
60  IV                   F88
61  K                    E86
62  MS.REPLICATION       G44
63  MS.RESIDUAL          G46
64  MS.TREATMENT         G45
65  N                    D76
66  SS.REPLICATION       E44
67  SS.RESIDUAL          E46
68  SS.TOTAL             E47
69  SS.TREATMENT         E45
70
71
72  4.A.   Sums and counts
73
74  SUM(W)       124.0339   88.6911  116.7086
75  SUM(W)^2    15384.408 7866.1112 13620.897
76  COUNT(W)           20        20        20
77
78  4.B.   Computing table for randomized block
79  (from Koch and Link, 1970, p. 159)
80
```

Worksheet C4D *(continued)*

```
81    ---------------------------------------------------------
82     Type       Total of Number of   Squares/  Element
83     of sum     squares    observ.    observ.
84    ---------------------------------------------------------
85   Grand       108526.49         60 1808.7749        (I)
86   Replicat. 11894.494          3 3964.8316       (II)
87   Treat.      36871.416         20 1843.5708      (III)
88   Observ.    5140.8519          1 5140.8519       (IV)
89    ---------------------------------------------------------
90
91
92
```

Worksheet C4D *(continued)*

93 4-C. Squares of observations

Row	C	D	E	F	G	H	I
97	Squared				Group Number		
98	sample						
99	totals	1	2	3	4	5	6
101	566.273	27.019	39.113	152.387			
102	29.218	0.106	4.548	8.686			
103	15.719	0.008	0.000	15.027			
104	0.135	0.055	0.000	0.017			
105	13.672	0.000	0.000	13.657			
106	1584.295	573.890	78.473	48.842			
107	12.597	6.168	0.142	0.475			
108	0.938	0.184	0.112	0.042			
109	0.773	0.000	0.046	0.440			
110	131.084	69.076	0.799	5.036			
111	5.001	0.038	0.148	2.744			
112	358.352	18.481	41.270	67.355			
113	22.574	0.416	1.788	7.667			
114	2089.578	1.710	294.472	742.236			
115	1434.743	526.450	60.513	51.187			
116	670.044	360.134	10.138	13.868			
117	3403.311	128.068	652.583	461.197			
118	5.988	0.692	0.539	0.776			
119	0.000	0.000	0.000	0.000			
120	1550.202	505.764	50.030	96.240			

CHAPTER 5

Exploration Geochemical Data

1-2-3 provides useful procedures for analyzing sets of exploration geochemical data that are not too large. (The size will depend on your computer memory; a maximum typical size may be a few hundred sites with ten to twenty variables per site.) This chapter applies some of these procedures to a small set of geochemical data.

For larger data sets, it is more effective to use a system of BASIC programs (Koch, 1987), or one of the general purpose programs for scientific data analysis, for example, MINITAB (Ryan, Joiner, and Ryan, 1985). Fortunately, it is not difficult to transfer files from 1-2-3 to other programs (Koch, 1987, p. 16) or from other programs to 1-2-3.

5.1 SUMMARY STATISTICS

Worksheet C5A.WK1 calculates summary statistics for a Norwegian data set, which Howarth and Sinding-Larsen (1983) used to demonstrate multivariate statistical analysis. Part 1 is their data set; Part 2 gives basic statistics, calculated using 1-2-3 functions (except for CV, the coefficient of variation, which is the ratio of the standard deviation, STD, to the arithmetic average or sample mean, AVG). As with worksheet C4A.WK1, in order to avoid resetting the formulae in Part 2, you can increase the number of rows using the 1-2-3 command /Worksheet Insert Row, and the number of columns using the command /Copy.

5.2 FREQUENCY DISTRIBUTIONS AND HISTOGRAMS

Worksheet C5B.WK1 calculates frequency distributions for the values of Zn and Fe in worksheet C5A.WK1. Part 1 is the same as Part 1 of the previous worksheet. Part 2 provides the frequency distributions, obtained using 1-2-3's /Data Distribution command. Here are the step-by-step directions for calculating the frequency distribution for zinc.

(1) In range C54..C78, establish the class intervals (named "bins" by 1-2-3) using the /Data Fill command. I used a starting value of 0 and a step value of 50.

(2) Using the /Data Distribution command, calculate the frequencies. 1-2-3 first asks for the range of values, which is D15..D39, and then for the range of bins, established in the previous step as C54..C78.

(3) Make a histogram for zinc (Figure 5.1) by using the graphics commands, as follows:

(a) Select /Graph Type and enter B for a bar graph.

(b) Select X and specify the range C54..C75 for the range of values on the x-axis.

(c) Select A and specify the range D54..D75 for the range of values on the y-axis.

(d) Select Options Titles and specify titles for the two title lines and the x- and y-axes.

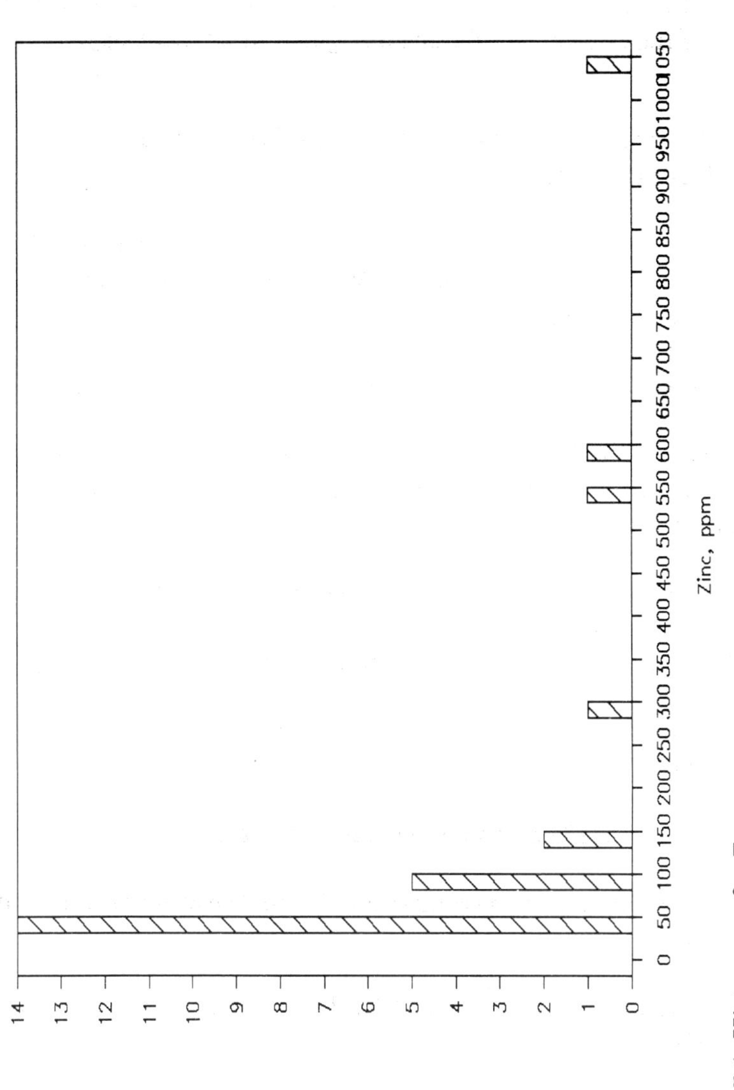

Figure 5.1 Histogram for Zn, Norwegian data set

(e) Select View to display the graph on the screen.

Repeating the analysis for Fe gives the results in the worksheet and in Figure 5.2.

1-2-3's /Data Distribution command does not provide some information that is conventional for frequency distributions, and the graphics commands display bar graphs rather than usual histograms with the bars adjacent to one another. In Part 3 of the worksheet, I have recast the frequency distribution for Zn in a conventional form, through a few additional steps. You can make a conventional histogram with 1-2-3's / Graph Type XY command by specifying the coordinates of the corners of the bars for the X and A ranges. Doing this is simple but tedious, and seems to me to defeat 1-2-3's purpose of providing graphs that are easy to both make and read. Although 1-2-3 macros could be written to handle the tasks explained in this paragraph, I would prefer to use another computing method altogether.

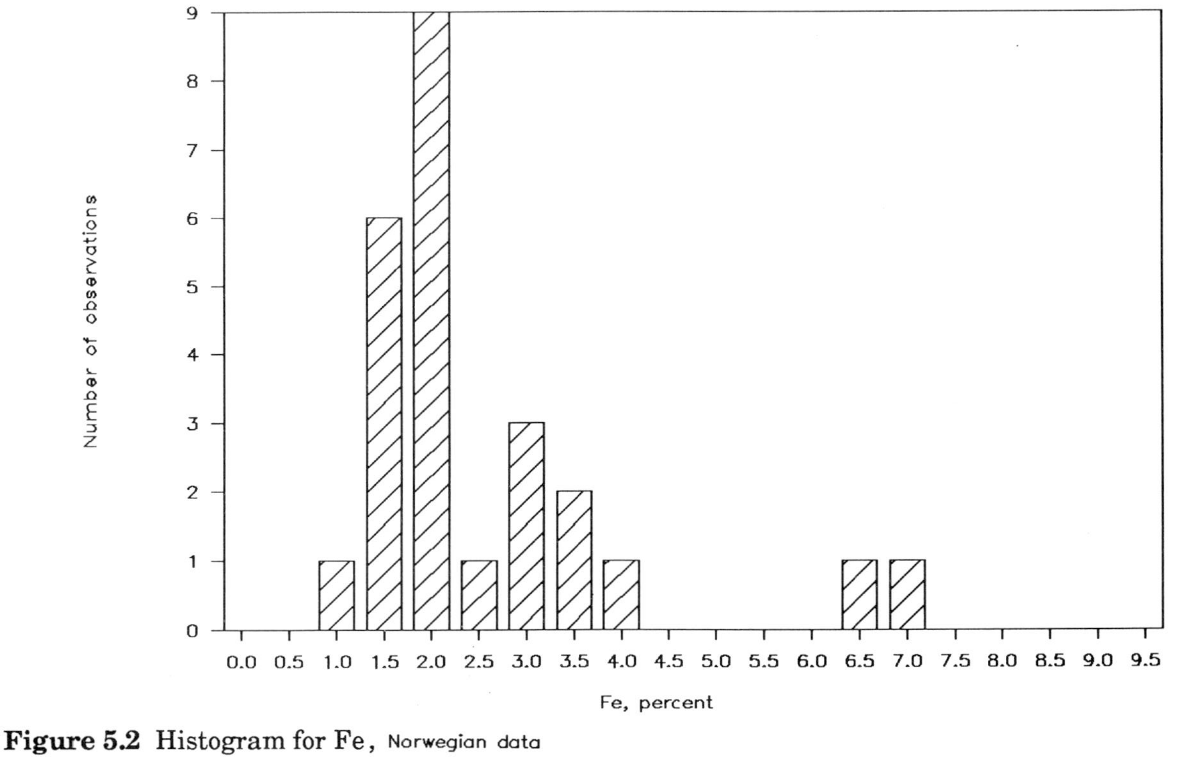

Figure 5.2 Histogram for Fe, Norwegian data

Worksheet C5A Basic statistics for exploration geochemical data

1 28 Aug 87, file C5A.WK1 on disks GSK 059-061
2
3 Worksheet C5A.WK1. Basic statistics for
4 exploration geochemical data
5
6
7 Part 1. Table of stream-sediment data from Norway
8 (from Howarth and Sinding-Larsen, 1983, p. 210) 9

Row	C	D	E	F	G	H	I
12	Sample	Zn	Fe	Mn	Cd	Cu	Pb
13	No.	ppm	pct	ppm	ppm	ppm	ppm
15	1	24	1.08	330	0.4	7	5
16	2	25	1.18	420	0.3	9	7
17	3	42	2.06	910	0.6	12	6
18	4	50	1.73	700	0.5	15	9
19	5	52	1.74	690	0.5	18	10
20	6	29	1.06	510	0.6	10	8
21	7	26	1.08	530	0.5	10	7
22	8	23	0.93	260	0.4	7	7
23	9	89	1.84	670	0.9	32	6
24	10	72	3.35	530	0.5	11	8
25	11	31	1.51	350	0.4	11	4
26	12	115	1.59	650	1.2	37	6
27	13	535	2.59	960	3.5	350	9
28	14	48	1.71	570	0.3	17	9
29	15	1010	2.71	1070	5.6	590	9

Worksheet C5A *(continued)*

30	16	560	3.05	450	3.1	490	8
31	17	48	1.43	590	0.4	8	8
32	18	44	1.70	710	0.4	11	16
33	19	36	1.28	410	0.2	7	10
34	20	33	1.94	490	1.0	20	12
35	21	45	1.79	260	0.8	27	12
36	22	118	6.70	1930	1.2	33	17
37	23	274	6.10	5920	3.1	63	27
38	24	81	2.62	970	0.9	22	13
39	25	80	3.61	900	1.0	23	10
40	------	------	------	------	------	------	------
41							
42							

Worksheet C5A *(continued)*

```
43   Part 2.  Basic statistics for all samples
44
45      C          D            E            F            G            H            I
46   -----------------------------------------------------------------------------------
47   Statistic    Zn           Fe           Mn           Cd           Cu           Pb
48                ppm          pct          ppm          ppm          ppm          ppm
49   -----------------------------------------------------------------------------------
50   N                 25           25           25           25           25           25
51   AVG              140         2.26          871          1.1           74           10
52   VAR            53464         2.09      1226344          1.7        24446           23
53   STD              231         1.44         1107          1.3          156            5
54   CV              1.66         0.64         1.27         1.15         2.12         0.49
55   MAX             1010         6.70         5920          5.6          590           27
56   MIN               23         0.93          260          0.2            7            4
57   -----------------------------------------------------------------------------------
58   N, number of observations; AVG, average (mean); VAR, variance;
59   STD, standard deviation; CV, coefficient of variation;
60   MAX, maximum value; MIN, minimum value.
61
62
63   Part 3.  Table of labels
64
65   AVG        D51
66   CV         D54
67   MAX        D55
68   MIN        D56
69   N          D50
70   STD        D53
71   VAR        D52
```

Worksheet C5B Frequency distributions and histograms for exploration geochemcial data

```
 1   10 Sep 87, file C5B.WK1 on disks GSK 059-061
 2
 3   Worksheet C5B.WK1.  Frequency distributions and histograms
 4       for exploration geochemical data
 5
 6
 7   Part 1.  Table of stream-sediment data from Norway
 8     (from Howarth and Sinding-Larsen, 1983, p. 210)
 9
10      C         D         E         F         G         H         I
11    ---------------------------------------------------------------------
12    Sample      Zn        Fe        Mn        Cd        Cu        Pb
13     No.        ppm       pct       ppm       ppm       ppm       ppm
14    ---------------------------------------------------------------------
15          1        24      1.08       330       0.4         7         5
16          2        25      1.18       420       0.3         9         7
17          3        42      2.06       910       0.6        12         6
18          4        50      1.73       700       0.5        15         9
19          5        52      1.74       690       0.5        18        10
20          6        29      1.06       510       0.6        10         8
21          7        26      1.08       530       0.5        10         7
22          8        23      0.93       260       0.4         7         7
23          9        89      1.84       670       0.9        32         6
24         10        72      3.35       530       0.5        11         8
25         11        31      1.51       350       0.4        11         4
26         12       115      1.59       650       1.2        37         6
27         13       535      2.59       960       3.5       350         9
28         14        48      1.71       570       0.3        17         9
29         15      1010      2.71      1070       5.6       590         9
```

Worksheet C5B *(continued)*

```
30        16       560     3.05      450     3.1     490      8
31        17        48     1.43      590     0.4       8      8
32        18        44     1.70      710     0.4      11     16
33        19        36     1.28      410     0.2       7     10
34        20        33     1.94      490     1.0      20     12
35        21        45     1.79      260     0.8      27     12
36        22       118     6.70     1930     1.2      33     17
37        23       274     6.10     5920     3.1      63     27
38        24        81     2.62      970     0.9      22     13
39        25        80     3.61      900     1.0      23     10
40   -------------------------------------------------------------
41
42
43  Part 2.  Frequency distributions for Zn and Fe
44
45     C        D         E         F         G         H         I
46   -------------------------------------------------------------
47          Zn                              Fe
48  -----------------              -----------------
49    Class                          Class
50  Interval                       Interval
51   (lower                          (lower
52   bound)  Frequency               bound)  Frequency
53   -----------------------------------------------------
54         0        0                   0.0        0
55        50       14                   0.5        0
56       100        5                   1.0        1
57       150        2                   1.5        6
58       200        0                   2.0        9
```

Worksheet C5B *(continued)*

59	250	0	2.5	1
60	300	1	3.0	3
61	350	0	3.5	2
62	400	0	4.0	1
63	450	0	4.5	0
64	500	0	5.0	0
65	550	1	5.5	0
66	600	1	6.0	0
67	650	0	6.5	1
68	700	0	7.0	1
69	750	0	7.5	0
70	800	0	8.0	0
71	850	0	8.5	0
72	900	0	9.0	0
73	950	0	9.5	0
74	1000	0		0
75	1050	1		
76	1100	0		
77	1150	0		
78	1200	0		
79		0		
80	----------	----------	----------	----------
81				
82				

Worksheet C5B *(continued)*

83 Part 3. Frequency distribution for Zn

Row	C	D	E	F	G	H	I
84							
87–90	Class Interval		Frequency	Relative Freq. (%)	Cumula-tive Freq.	Relative Cumula-tive Freq. (%)	
92	0	50	0	0	0	0	
93	50	100	14	56	14	56	
94	100	150	5	20	19	76	
95	150	200	2	8	21	84	
96	200	250	0	0	21	84	
97	250	300	0	0	21	84	
98	300	350	1	4	22	88	
99	350	400	0	0	22	88	
100	400	450	0	0	22	88	
101	450	500	0	0	22	88	
102	500	550	0	0	22	88	
103	550	600	1	4	23	92	
104	600	650	1	4	24	96	
105	650	700	0	0	24	96	
106	700	750	0	0	24	96	
107	750	800	0	0	24	96	
108	800	850	0	0	24	96	
109	850	900	0	0	24	96	
110	900	950	0	0	24	96	
111	950	1000	0	0	24	96	

Worksheet C5B *(continued)*

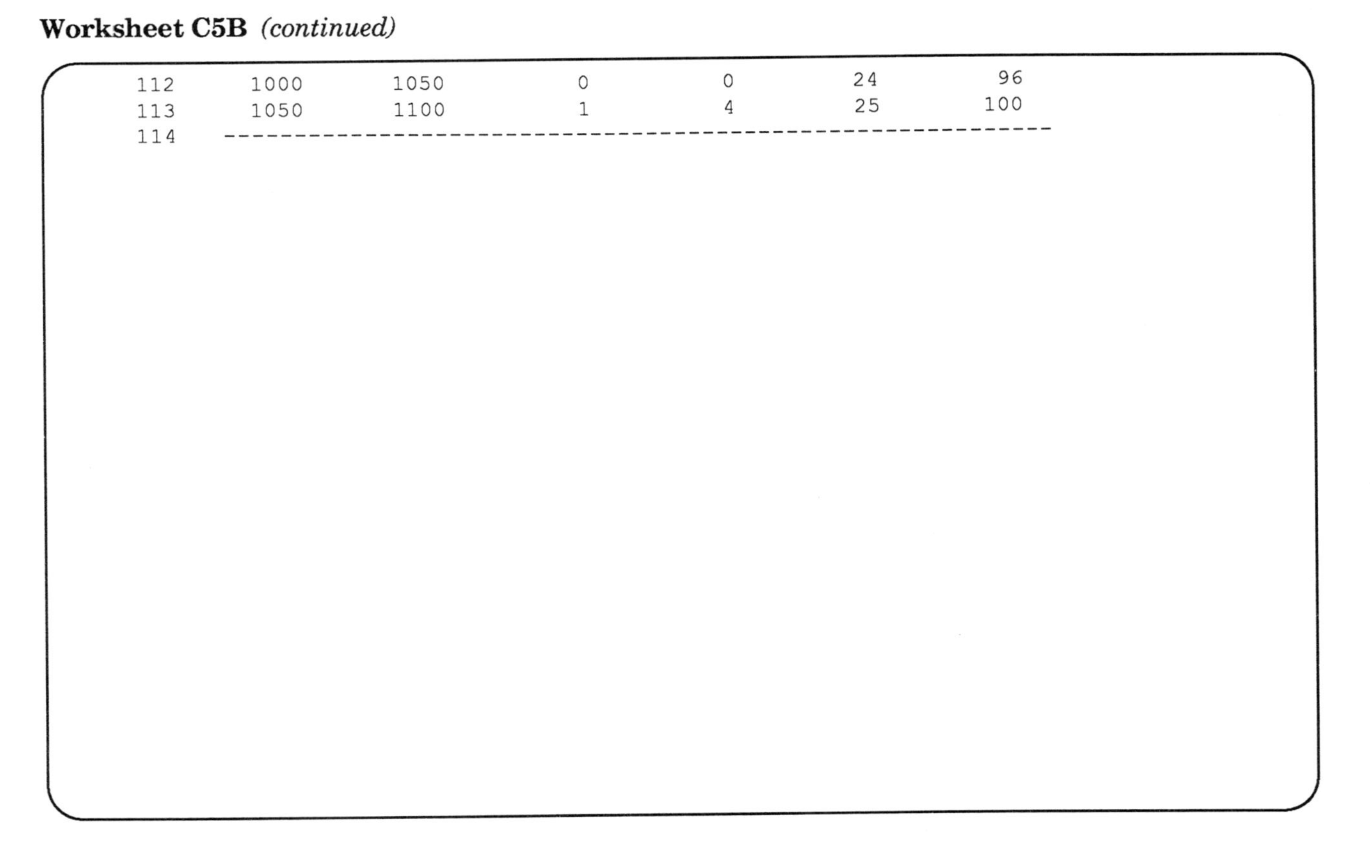

```
112   1000   1050   0   0   24    96
113   1050   1100   1   4   25   100
114   ------------------------------------
```

CHAPTER 6

Exploration Models

This chapter will discuss two simulation models for drilling on grids; for simplicity we will consider only square grids. After practicing with the worksheets, you will be able to modify them for other situations or write additional ones. The worksheets in this chapter and the next contain "macros", which are short programs written in 1-2-3's command language. Macros make it easy to perform certain operations.

6.1 A SIMULATION OF GRID DRILLING

In mineral exploration, grid drilling is used widely, particularly for the discovery and evaluation ofbedded deposits including uranium, coal, kaolin, and many others. In an area to be explored, if we can estimate the percentage of total area that is mineralized and assume that the mineralized areas are "randomly distributed," we can then simulate the effect of drilling holes to discover ore deposits. "Randomly distributed" indicates that any part of the area has an equal chance of containing an orebody as any other. Simulation is "a class of techniques that involve setting up a model of a real situation and then performing experiments on the model," to quote from the introduction to a comprehensive and classic book on simulation in geology (Harbaugh and Bonham-Carter,

1970, p. 1-2). The ideas in this chapter are drawn from a paper by Koch and Schuenemeyer (1982). First, I will explain the worksheet and then suggest some modifications that you may want to make.

Worksheet C6A.WK1 simulates the drilling of vertical holes on a square grid of 100 potential locations in a 10 by 10 array. After you specify drillhole locations in these 100 potential locations, the worksheet indicates which of the "drillholes" encountered ore. To run the model, execute the four steps summarized at the beginning of the worksheet in any order that you select.

Part 1 of the worksheet is an abbreviated set of instructions for the worksheet's four steps. Step 1 allows you to specify the percent area that is mineralized. To do this, press ALT-a (this notation indicates to press the ALT key and then the a key, while holding down the ALT key); this sequence of keys calls a 1-2-3 command named a "Macro," used in 1-2-3 for many purposes). On the second line of the screen, the following request appears:

Enter percent area mineralized:

In response, enter any appropriate percentage. For the worksheet on the diskette, I entered 10. (The worksheet stores this percentage in cell C105 and the corresponding decimal fraction in cell C106.

In Step 2, first place cell A21 at the upper-left-hand corner of your screen, and then enter drillhole locations in the array MATRIX.HOLES (Part 2) in cells C21..L30. These locations can be identified by any numerical values; for illustration, I entered the numbers 1 to 10. The hits then appear as asterisks in the array MATRIX.HITS in cells C31..L40. Matrices ARRAY.HOLES and ARRAY.HITS are formatted so that they appear on one screen through your placement of cell A21.

How does the program determine whether a drillhole hits ore? This is done by comparing the matrix of hole locations, MATRIX.HOLES, with MATRIX.DATA (Part 3), which is a matrix (in cells C51..L60), of random numbers between 0 and 1. If the random number is low enough, relative to the percent area selected as mineralized in Step 1, a hit is recorded. For the example, only hole 5 in cell F25 is a hit, because only the random number 0.010 in cell F55 is less than 0.1.

Step 3 allows you to erase one set of hole locations and start over by using the macro ALT-b.

Step 4 allows you to select a different set of random numbers for array MATRIX.DATA by using the macro ALT-c, which copies the random numbers in array MATRIX.RAND (Part 4) into array MATRIX.DATA. The array MATRIX.RAND is reset each time you press function key 9.

Here are some notes about this model. In order to save room on the diskette, I used a 10 by 10 grid. 1-2-3 will allow you to make the grid larger if you first copy the worksheet to a diskette used for only this model. Although the model is for holes on a square grid, the computer monitor display is not square in order to simplify programming. Using the command /Worksheet Global Column-width, you can reformat the cell size if you wish. Also, the model can readily be modified for non-square grids. And it would not be difficult to introduce a gradient into the random numbers, corresponding to a change in mineralization favorability in some direction. For example, this change would be appropriate for a mine in the Coeur d'Alene district (Koch, Schuenemeyer, and Link, 1974). Taking account of the probabilities of recognition given a hit also would be interesting (Koch and Schuenemeyer, 1982, p. 654).

6.2 THE ELMWOOD, TENNESSEE ZINC DEPOSIT

Between 1964 and about 1980, more than 20 mining companies explored the Middle Tennessee district near Nashville for zinc deposits. These deposits are located at depths of between 300 and 600 meters in gently dipping rocks of Ordovician age. Because the ore deposits occur below a major unconformity which is not exposed in the district, surface geology provided little or no guide to ore; nor were geophysics or geochemistry of much help. Therefore, nearly all exploration was by diamond drilling.

Although active exploration was essentially stopped in the 1980's for economic reasons, the district contains a major zinc resource for the future.

In 1967, The New Jersey Zinc Company discovered the Elmwood mine near Carthage, Tennessee after drilling 79 holes; Callahan (1977) details the exploration campaign. After the discovery, the deposit was evaluated through drilling another 89 holes on a square grid with a hole spacing of 1000 feet.

Worksheet C6B.WK1 maps the zinc grades located by drilling at Elmwood. We can use these data for two simulation exercises: (1) to simulate the discovery process, and (2) to define a mining area, based on the entire data set. These two exercises are taken up in turn.

Part 1 is instructions for the four macros that run the model. Part 2 is array MATRIX.DATA, which is an 11 by 11 array, consisting of Callahan's (1977, p. 1390) 89 holes, plus another 17 holes drilled by the New Jersey Zinc Company, plus 15 holes simulated by me to complete the grid. (Callahan's original data are given in array ORIGINAL.DATA, displayed in Part 7.) When you retrieve the worksheet, the values in array MATRIX.DATA are not displayed (because macro \0 runs automatically), so that you can "discover" them by "drilling" holes. To see the values, press Alt-c; to hide them once more, press Alt-d.

To "drillholes," press Alt-a, and follow the instructions for entering holes according to the rows and columns of array MATRIX.HOLES, in Part 3. The rows and columns are numbered starting in the upper-left-hand corner according to 1-2-3's system. For the worksheet provided on the diskette, two holes are drilled, in row 2, column 3, and in row 4, column 5. The results of the drilling simulation appear in Part 4 in array MATRIX.HITS. For the example, the values are 2 and 5 percent zinc.

You then can continue the simulation to get an average grade for all the holes drilled or for selected holes by pressing Alt-b. This macro selects holes by cell addresses. Selecting addresses F70 and H72 gives the average grade of 3.5 percent zinc.

For the second simulation, you can display the entire array MATRIX.DATA by pressing Alt-c, and then proceed as noted. The assumption is that the entire grid has been drilled.

Part 5 lists the macros, and Part 6 is the table of labels. Part 7 is Callahan's original data matrix. To use this original matrix in your simulations, press Alt-e, which copies it to array MATRIX.DATA. Then, if you drill a hole where there is an empty cell, the worksheet records an o (lower-case "oh").

Here are some notes on this model. To scale the arrays for printing, I reduced the column-widths to 2, with the result that their labels are abbreviated. In the array MATRIX.DATA, row and column labels 0 indicates 10, and in the second set of labels, 1 indicates 11. In this matrix, the rows and columns start in the upper-left- hand corner, following 1-2-3's convention; in the convention usual in geology and engineering, the array is in the southeast or fourth quadrant. The IF statement in cell D69 is:

@IF(@ISSTRING(D49=0," ",@IF(D31)>0,D31,"o")).

The @ISSTRING function tests for a blank cell in array MATRIX.HOLES; if blank, a blank is placed in the corresponding cell in MATRIX.HITS. Otherwise, the @IF function tests for a value greater than 0 in array MATRIX.DATA; if greater, the value is recorded in the corresponding cell in MATRIX.HITS; if smaller, which happens if you use array ORIGINAL.MATRIX, an o is recorded to indicate no zinc. The greater-than-0 value can be readily changed for other simulations.

Worksheet C6A Simulation of drilling exploration holes

```
1   02 Feb 89, file C6A.WK1 on disks GSK 059-061 2
3   Worksheet C6A.WK1. Simulation of drilling exploration holes
4
5
6   Part 1. Abbreviated running instructions (Details in text.)
7
8   (1)   To specify the percent area mineralized, press Alt-a.
9
10   (2)   Specify drill-hole locations by entering numbers in array
11   MATRIX.HOLES, C21..L30.  The hit locations will appear in array
12   MATRIX.HITS, C31..L40.
13
14   (3)   To erase a set of drill-hole locations, press Alt-b.
15
16   (4)   To reset the mineralized cells, press Alt-c.
17
18
```

Worksheet C6A *(continued)*

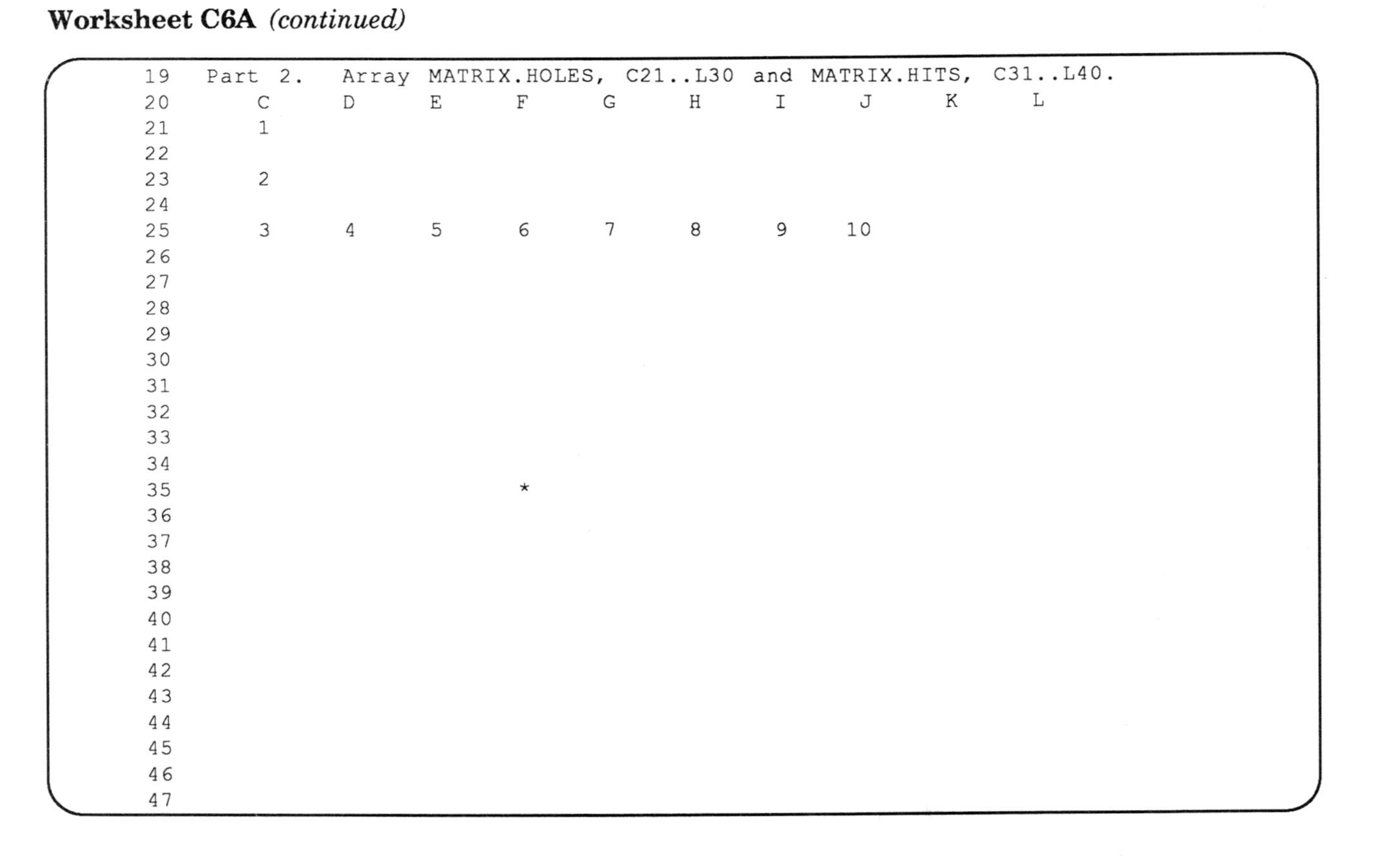

```
19   Part 2.   Array MATRIX.HOLES, C21..L30 and MATRIX.HITS, C31..L40.
20      C       D       E       F       G       H       I       J       K       L
21      1
22
23      2
24
25      3       4       5       6       7       8       9      10
26
27
28
29
30
31
32
33
34
35                              *
36
37
38
39
40
41
42
43
44
45
46
47
```

Worksheet C6A *(continued)*

```
48
49   Part 3.  Array MATRIX.DATA, C51..L60, of values between 0 and 1.
50
51   0.572 0.603 0.051 0.814 0.835 0.943 0.709 0.486 0.209 0.327
52   0.403 0.532 0.019 0.443 0.849 0.576 0.621 0.575 0.676 0.005
53   0.909 0.287 0.084 0.558 0.855 0.778 0.181 0.080 0.459 0.197
54   0.996 0.858 0.699 0.483 0.403 0.209 0.357 0.499 0.432 0.249
55   0.811 0.536 0.704 0.010 0.179 0.459 0.987 0.324 0.396 0.065
56   0.082 0.155 0.106 0.427 0.552 0.287 0.462 0.289 0.002 0.559
57   0.204 0.800 0.918 0.048 0.624 0.132 0.281 0.889 0.265 0.790
58   0.448 0.639 0.482 0.647 0.041 0.378 0.209 0.010 0.810 0.937
59   0.552 0.261 0.725 0.961 0.057 0.868 0.768 0.700 0.771 0.823
60   0.166 0.716 0.454 0.781 0.044 0.919 0.549 0.475 0.735 0.616
61
62
63   Part 4. Array MATRIX.RAND, C66..L75, of random numbers
64   between 0 and 1.
65
66   0.765 0.188 0.235 0.477 0.397 0.325 0.533 0.606 0.725 0.395
67   0.798 0.129 0.113 0.939 0.663 0.976 0.661 0.211 0.091 0.781
68   0.528 0.644 0.550 0.058 0.269 0.414 0.486 0.098 0.817 0.641
69   0.332 0.232 0.952 0.020 0.532 0.410 0.373 0.937 0.313 0.241
70   0.732 0.365 0.195 0.940 0.241 0.081 0.472 0.828 0.710 0.007
71   0.146 0.340 0.119 0.720 0.045 0.673 0.099 0.428 0.122 0.540
72   0.913 0.961 0.157 0.653 0.813 0.563 0.347 0.911 0.867 0.551
73   0.530 0.409 0.626 0.125 0.012 0.740 0.243 0.869 0.243 0.225
74   0.884 0.761 0.821 0.015 0.571 0.722 0.557 0.225 0.786 0.004
75   0.596 0.528 0.462 0.969 0.807 0.878 0.225 0.049 0.952 0.284
76
```

Worksheet C6A *(continued)*

```
77
78   Part 5.   Table of labels
79
80   %.AREA.MINERAL      C102
81   FRACT.AREA.MIN.     D102
82   MATRIX.DATA         C51..L60
83   MATRIX.HITS         C31..L40
84   MATRIX.HOLES        C21..L30
85   MATRIX.RAND         C66..L75
86   \A                  D93..D94
87   \B                  D96..D97
88   \C                  D99
89
90
91   Part 6.   Macros
92
93   \a      {GETNUMBER "Enter percent area mineralized:",C102}
94           {CALC}
95
96   \b      /reMATRIX.HOLES~
97           {CALC}
98
99   \c      /rvMATRIX.RAND~MATRIX.DATA~
100
101
102       10    0.1 C105 = % area mineralized; D105 = decimal fraction
```

Worksheet C6B Simulation of drilling holes at Elmwood, Tennessee

```
1   02 Feb 89, file C6B.WK1 on disks GSK 059-061 2
3   Worksheet C6B.WK1. Simulation of drilling holes at
         Elmwood, Tennessee
5
6   Part 1.  Abbreviated running instructions ( Details in text.)
7
8   (1)   To "drill" holes, press Alt-a.
9
10   (2)   To calculate the table of summary statistics in Part 4,
11         press Alt-b.
12
13   (3)   To hide the array MATRIX.DATA, press Alt-d.
14
15   (4)   To display the array MATRIX.DATA, press Alt-c.
16
17
18
19
20
21
22
23
24
```

Worksheet C6B *(continued)*

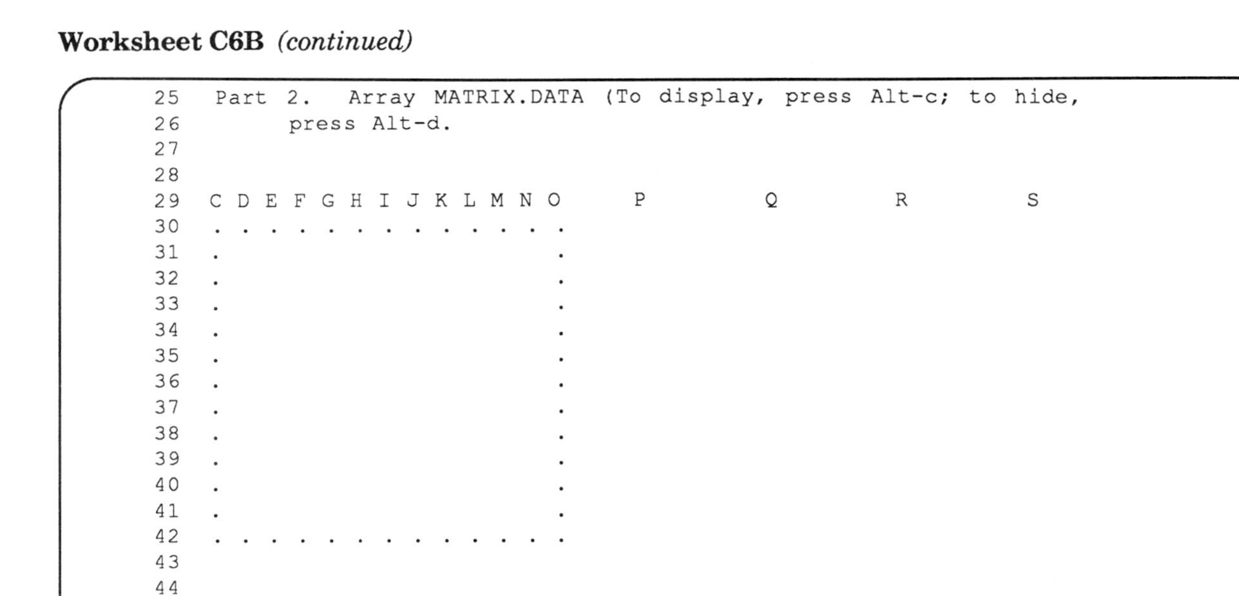

```
25  Part 2.   Array MATRIX.DATA (To display, press Alt-c; to hide,
26       press Alt-d.
27
28
29  C D E F G H I J K L M N O     P         Q         R         S
30  . . . . . . . . . . . . .
31  .                       .
32  .                       .
33  .                       .
34  .                       .
35  .                       .
36  .                       .
37  .                       .
38  .                       .
39  .                       .
40  .                       .
41  .                       .
42  . . . . . . . . . . . . .
43
44
```

Worksheet C6B *(continued)*

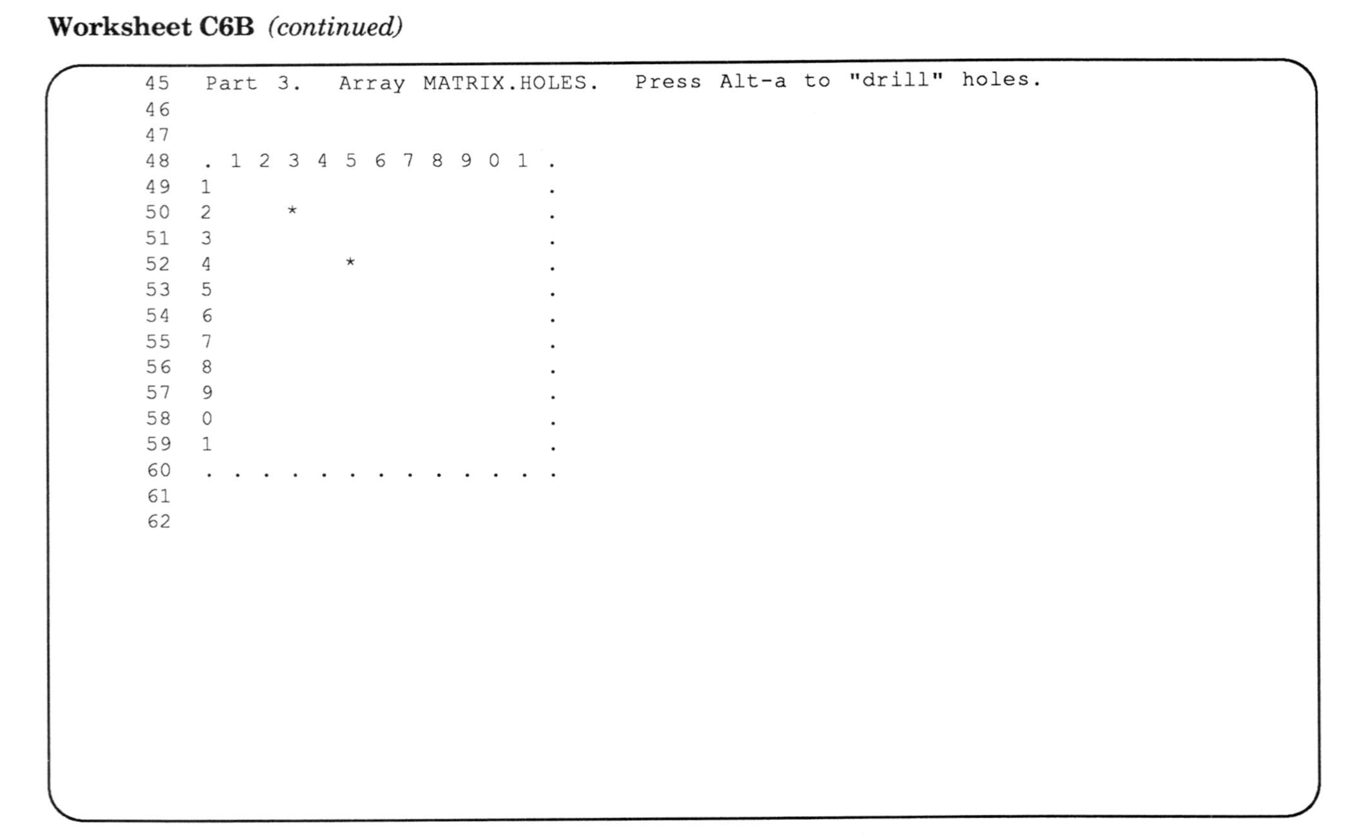

```
45  Part 3.   Array MATRIX.HOLES.   Press Alt-a to "drill" holes.
46
47
48  . 1 2 3 4 5 6 7 8 9 0 1 .
49  1                       .
50  2     *                 .
51  3                       .
52  4         *             .
53  5                       .
54  6                       .
55  7                       .
56  8                       .
57  9                       .
58  0                       .
59  1                       .
60  . . . . . . . . . . . . .
61
62
```

Worksheet C6B *(continued)*

```
63  Part 4.  Array MATRIX.HITS and Table of Summary Statistics
64       Press Alt-b to calculate the Table.
65
66
67    D E F G H I J K L M N
68  . . . . . . . . . . . . .        --------------------------
69  .                       .                  Hole
70  .     2                 .                  No.      Zn,%
71  .                        .       --------------------------
72  .         5             .      Hole           1          2
73  .                       .      Hole           2          5
74  .                       .      Hole           3
75  .                       .      Hole           4
76  .                       .      Hole           5
77  .                       .      Hole           6
78  .                       .      Hole           7
79  .                       .      Hole           8
80  . . . . . . . . . . . . .      Hole           9
81                                 Hole          10
82
83                                 No. of holes              2
84                                 Average grade, %        3.5
85
86                                 --------------------------
87
88
```

Worksheet C6B *(continued)*

```
89   Part 5. Macros
90                                  \a        /RED49..N59~
91                                            {GOTO}B45~
92                                             {GETNUMBER "Enter the numbe
93                                             {FOR COUNT1,1,NUM1,1,SUB1}
94
95                                  SUB1       {GETNUMBER "Enter the hole
96                                             {GETNUMBER "Enter the hole
97                                             {PUT MATRIX.HOLES,COL,ROW,"
98                                            {CALC}
99
100
101                                 COL                 5
102                                 COUNT1              3
103                                 NUM1                2
104                                 ROW                 4
105
106
107                                  \b        /RES72..S81~
108                                            {GOTO}B63~
109                                             {GETNUMBER "Enter the numbe
110                                            {DOWN 9}
111                                            {RIGHT 17}
112                                             {FOR COUNT,1,NUMBER,1,ROUTI
113                                            {CALC}
114                                            {GOTO}S81~
115                                            {DOWN 5}
116
117                                   ROUTINE  /XNEnter a drill-hole locat
```

Worksheet C6B *(continued)*

```
118                                                  {DOWN}
119
120
121                                  NUMBER              2
122                                  COUNT               3
123
124
125                                   \c          /RFGD31..N41~
126                                               {GOTO}B25~
127
128                                   \d          /RFHD31..N41~
129                                  \0
130
131                                   \e          /CD156..N166~D31..N41~
132                                               {GOTO}B25~
133
134  Part 6.   Table of labels
135
136  COL              Q101
137  COUNT            Q122
138  COUNT1           Q102
139  MATRIX.HOLES     C48..N59
140  NUM1             Q103
141  NUMBER           Q121
142  ROUTINE          Q116
143  ROW              Q104
144  SUB1             Q95
```

Worksheet C6B *(continued)*

```
145  \0               Q128
146  \A               Q90
147  \B               Q107
148  \C               Q125
149  \D               Q128
150  \E               Q131
151
152
153  Part 7.   Array ORIGINAL.DATA
154
155  . . . . . . . . . . . . .
156  .   1           1   1 3 .
157  . 1   2 2 3 3   1 1 1 5 .
158  . 1 1 5 1 1 3 3 1 2 1 1 .
159  .   1 2 2 5 1 1   3 1 2 .
160  . 1 1 1 1 2 3 1 1 1   1 .
161  . 1 1 1 3 1 5 5 1 3 1 5 .
162  . 3 1 5 5 1 5 1 1 1 1 2 .
163  . 1 1 1 1 1 5 1 1 1 1 1 .
164  . 2 1 2 1   3 1 1 2 2 1 .
165  .   1 1 1 1 3 5 2 1 1 1 .
166  . 1 1 1 1   2 2 5 1 3 1 .
167  . . . . . . . . . . . . .
```

CHAPTER 7

Probability and Related Calculations

This chapter explains some probability calculations used in exploration and also ways to plot tonnage-grade curves. The final section tells how to plot logarithmic graphs using 1-2-3.

7.1 TONNAGE-GRADE RELATIONS

Lasky (1950) proposed that the tonnage and grade of ore deposits follow an exponential relationship that plots as a straightline on semi-logarithmic graph paper. Koch and Link (1971, p. 255-262) summarize Lasky's work, and Harris (1984, p. 59-92) gives a comprehensive account of Lasky's and later investigations of this complex and controversial subject.

Worksheet C7A.WK1 provides the data to graph Lasky's law (Figure 7.1) with his constants for a hypothetical porphyry copper deposit. Lasky's formula appears in cells G32..G34, and the scaled logarithmic values in cells H32..H34. The previously cited references explain the mathematics, and the plotting procedure is explained in Section 7.3. You can readily modify this worksheet to plot graphs for other deposits using data from Lasky, Harris, or other workers.

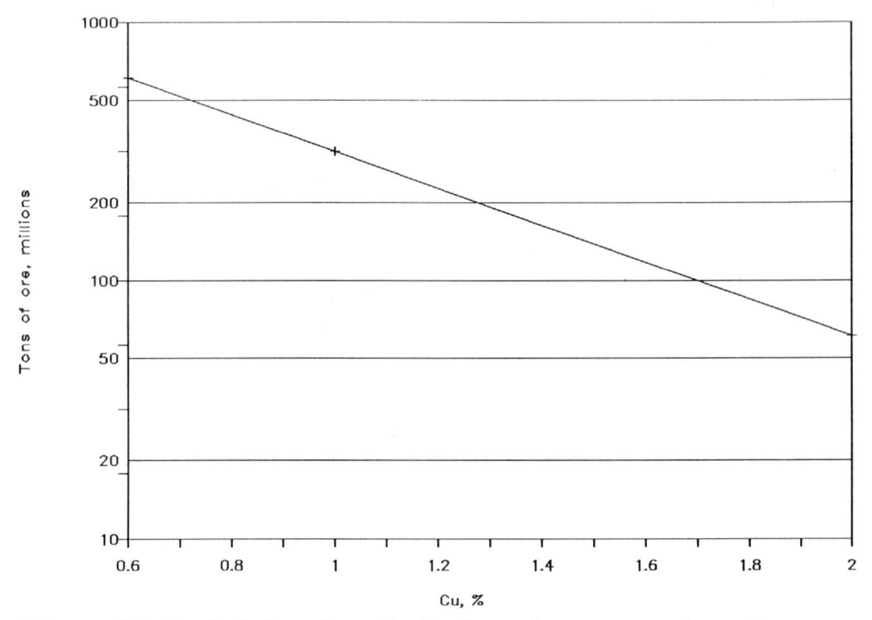

Figure 7.1 Lasky's law, hypothetical porphyry copper deposit

Worksheet C7B.WK1 provides the data to graph later work on copper deposits by Cox, Wright, and Coakley, (1981), who tabulate (Table 1, p. F3-F6) published reserves and resources of copper in the United States. In the worksheet, the tonnage data are listed in range C21..C94, and the grade data in range E21..E94.

Figure 7.2 plots these data for the entire United States, and Figure 7.3 plots the subset for operating mines and announced developments in Arizona. Using 1-2-3, it is easy to edit the data list to obtain a graph for any interesting subset, for instance, for all deposits in other states, or all deposits of a certain geologic type.

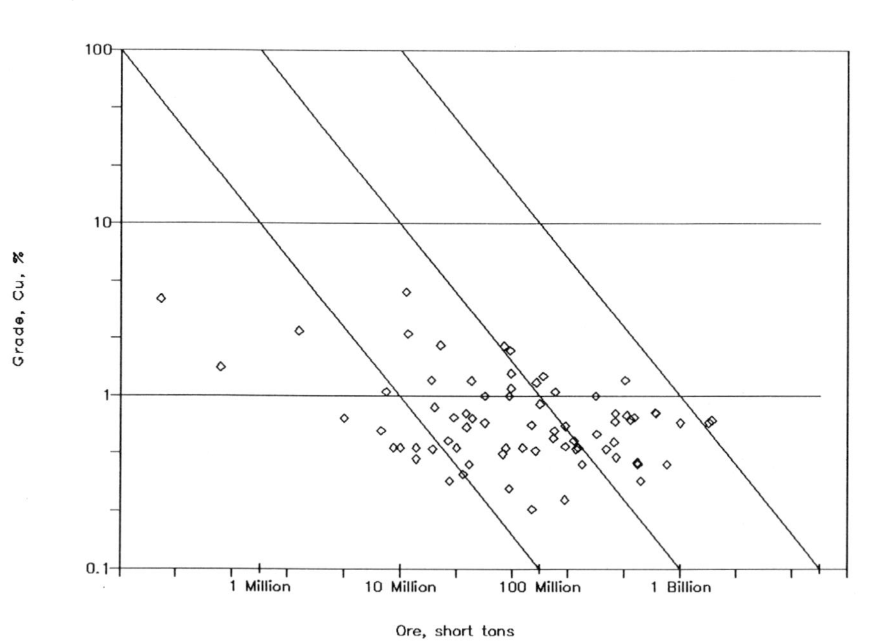

Figure 7.2 U.S. copper reserves, tonnages, and grades (from Cox, 1981)

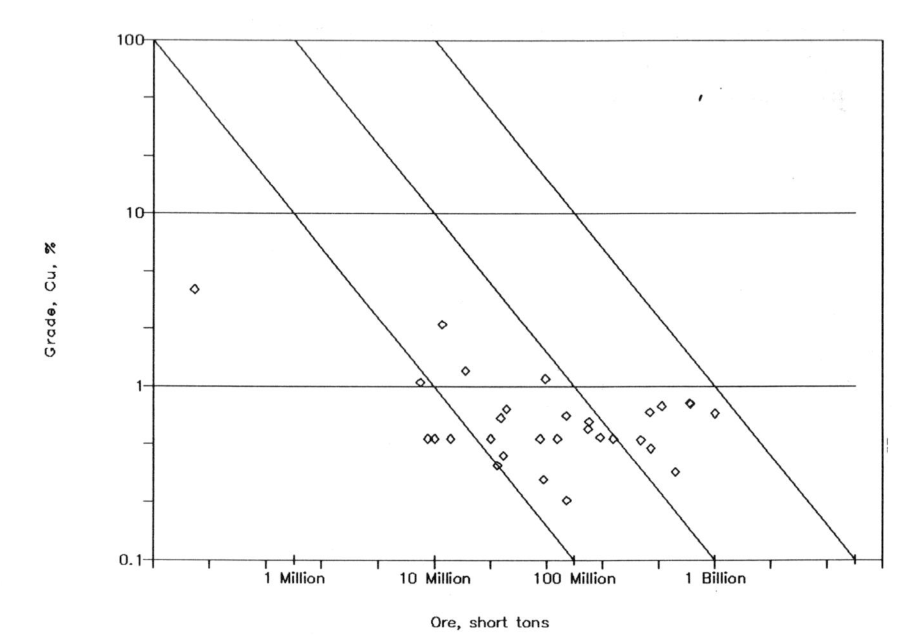

Figure 7.3 Arizona copper reserves , tonnages, and grades (from Cox, 1981)

7.2 PROBABILITY CALCULATIONS

Worksheet C7C.WK1 provides 1-2-3 macros to facilitate some of the probability calculations used in exploration. These calculations are explained for mineral deposits by Koch and Link (1971, p. 187-228) and Peters (1987, p. 444-448) and for petroleum exploration by Harbaugh, Doveton, and Davis (1977). There are too many approaches and formulae to include in this book. The worksheet provides macros for two simple situations, Gambler's ruin, and calculation of a factorial, needed for many probability formulae, but not furnished as a function by 1-2-3.

7.3 PLOTTING LOGARITHMIC GRAPHS

These directions for plotting logarithmic graphs are simplified from those of Fine (1987); they are given here because of the importance of logarithmic graphs in geology, and because Fine's article may not be readily available to you.

Because 1-2-3 does not provide for logarithmic plots, it is necessary to select the range of values to plot, calculate the logarithms of this range, plot the range, and relabel the range with the original values rather than the logarithms. None of these steps is complicated in itself, but combining them together in 1-2-3 takes some planning and patience. The principal point to remember is that 1-2-3's basic concept is that of the spreadsheet, or array, so that the data to be plotted need to be expressed as an array.

To explain logarithmic plotting, I will demonstrate how to plot Figure 7.1. In worksheet C7A, the X values are listed in range E11.E34, the A values in range F11..F30, and the B values in range H11..H34. 1-2-3 plots points only where X values are paired with A or B values. The A values define the boundaries of the graph and the horizontal lines; the B values define the data points.

This is the sequence of commands with their purposes:

(1) /Graph Type XY, to select the appropriate display.

(2) /Graph Options Scale Y Scale Format Hidden Quit Quit, to suppress automatic scaling on the Y axis.

(3) /Graph, select X, and specify range E11..E34.

(4) /Graph, select A, and specify range F11..F30.

(5) /Graph Options Format A Lines Quit Quit, to plot lines only.

(6) /Graph Options Data-Labels A (Specify range I11..I29) Left Quit Quit, to plot logarithmic values for the labels.

(7) /Graph, select B, and specify range H11..H34 to plot the data points. Even though the data points appear only in the range H32..H34, the entire range must be specified.

(8) /Graph Options Titles, First, and specify the first title line. Repeat for the second title line, and for the X- and Y-axis titles.

(9) Select View to display the graph on the screen.

Worksheet C5B Graph of Figure 7.1

1 15 Sep 87, file C7A.WK1 on disks GSK 059-061

2

3 Worksheet C7A.WK1. Graph of Figure 7.1

4

5

Row	C	D	E	F	G	H	I
6	C	D	E	F	G	H	I
7							
8		Scaled			Tonnage,		A
9	Y	LN	X	A	millions	B	Labels
10							
11	10	0.00	0.60	0.00			10
12	20	0.06	2.00	0.00			
13	50	0.14					
14	100	0.20	0.60	0.06			20
15	200	0.26	2.00	0.06			
16	500	0.34					
17	1000	0.40	0.60	0.14			50
18			2.00	0.14			
19							
20			0.60	0.20			100
21			2.00	0.20			
22							
23			0.60	0.26			200
24			2.00	0.26			
25							
26			0.60	0.34			500
27			2.00	0.34			
28							
29			0.60	0.40			1000

Worksheet C7A *(continued)*

```
30                          2.00     0.40
31
32                          0.60                610.54   0.3571
33                          1.00                316.23   0.3000
34                          2.00                 61.05   0.1571
35     ------------------------------------------------------------------
```

Worksheet C7B Graph of Figures 7.2 and 7.3

```
 1   15 Sep 87, file C7B.WK1 on disks GSK 059-061
 2
 3   Worksheet C7B.WK1. Graph of Figures 7.2 and 7.3
 4
 5
 6      C      D       E      F      G      H      I       J       K       L
 7   ------------------------------------------------------------------------
 8          Scaled         Scaled                            B       A
 9      X     LN       Y     LN      X      A      B   Labels  Labels     C
10   ------------------------------------------------------------------------
11      0.1  0.00    0.10  0.00   0.00    0.0                      0.1
12      1.0  0.50    1.00  0.30   2.50    0.0
13     10.0  1.00   10.00  0.60
14    100.0  1.50  100.00  0.90   0.00    0.3                        1
15     1000  2.00                 2.50    0.3
16    10000  2.50
17                                0.00    0.6                       10
18   ---------------------------  2.50    0.6
19          Cox Data
20   ---------------------------  0.00    0.9                      100
21    344.0  1.77    0.71  0.26   2.50    0.9
22     62.0  1.40    1.11  0.31
23     87.1  1.47    0.68  0.25   0.00    0.9
24     32.7  1.26    0.74  0.26   1.50    0.0
25     16.7  1.11    1.23  0.33
26    152.1  1.59    0.51  0.21   0.50    0.9
27      7.9  0.95    1.06  0.31   2.00    0.0
28     29.7  1.24    0.66  0.25
29    125.0  1.55    0.57  0.23   1.00    0.9
```

Worksheet C7B *(continued)*

30	350.0	1.77	0.44	0.19	2.50	0.0			
31	11.3	1.03	2.28	0.41					
32	297.0	1.74	0.49	0.21	0.00		0		
33	28.0	1.22	0.35	0.16	0.50		0	1 Million	
34	88.0	1.47	0.22	0.10	1.00		0	10 Million	
35	0.2	0.15	3.65	0.47	1.50		0	100 Million	
36	13.0	1.06	0.50	0.21	2.00		0	1 Billion	
37	31.0	1.25	0.40	0.18	2.50		0		
38	523.0	1.86	0.32	0.15					
39	60.0	1.39	0.29	0.14	1.77				0.26
40	189.0	1.64	0.50	0.21	1.40				0.31
41	25.2	1.20	0.50	0.21	1.47				0.25
42	56.8	1.38	0.50	0.21	1.26				0.26
43	9.0	0.98	0.50	0.21	1.11				0.33
44	667.0	1.91	0.79	0.27	1.59				0.21
45	1000.0	2.00	0.70	0.25	0.95				0.31
46	10.0	1.00	0.50	0.21	1.24				0.25
47	662.5	1.91	0.80	0.27	1.55				0.23
48	416.0	1.81	0.77	0.27	1.77				0.19
49	126.6	1.55	0.63	0.24	1.03				0.41
50	75.0	1.44	0.50	0.21	1.74				0.21
51	405.0	1.80	1.23	0.33	1.22				0.16
52	94.0	1.49	1.20	0.32	1.47				0.10
53	128.0	1.55	1.06	0.31	0.15				0.47
54	152.0	1.59	0.67	0.25	1.06				0.21
55	17.0	1.12	0.49	0.21	1.25				0.18
56	253.0	1.70	0.60	0.23	1.86				0.15
57	7.3	0.93	0.63	0.24	1.39				0.14
58	29.3	1.23	0.79	0.27	1.64				0.21

Worksheet C7B *(continued)*

59	443.0	1.82	0.73	0.26	1.20	0.21
60	344.4	1.77	0.79	0.27	1.38	0.21
61	19.3	1.14	1.97	0.39	0.98	0.21
62	17.5	1.12	0.86	0.28	1.91	0.27
63	40.0	1.30	1.00	0.30	2.00	0.25
64	61.2	1.39	1.84	0.38	1.00	0.21
65	1602.0	2.10	0.70	0.25	1.91	0.27
66	22.0	1.17	0.55	0.22	1.81	0.27
67	337.0	1.76	0.54	0.22	1.55	0.24
68	24.0	1.19	0.75	0.26	1.44	0.21
69	470.0	1.84	0.75	0.26	1.80	0.33
70	55.0	1.37	1.95	0.39	1.49	0.32
71	181.0	1.63	0.49	0.21	1.55	0.31
72	250.0	1.70	1.00	0.30	1.59	0.25
73	800.0	1.95	0.40	0.18	1.12	0.21
74	200.0	1.65	0.40	0.18	1.70	0.23
75	40.0	1.30	0.70	0.25	0.93	0.24
76	175.0	1.62	0.55	0.22	1.23	0.27
77	4.0	0.80	0.74	0.26	1.82	0.26
78	0.5	0.36	1.46	0.35	1.77	0.27
79	105.0	1.51	1.30	0.33	1.14	0.39
80	62.0	1.40	1.35	0.34	1.12	0.28
81	100.0	1.50	0.90	0.29	1.30	0.30
82	100.0	1.50	0.90	0.29	1.39	0.38
83	1679.0	2.11	0.72	0.26	2.10	0.25
84	93.0	1.48	0.48	0.20	1.17	0.22
85	151.0	1.59	0.25	0.12	1.76	0.22
86	1.9	0.64	2.37	0.41	1.19	0.26
87	22.4	1.18	0.32	0.15	1.84	0.26

Worksheet C7B *(continued)*

88	13.0	1.06	0.43	0.19	1.37	0.39
89	495.0	1.85	0.41	0.18	1.63	0.21
90	32.0	1.25	1.22	0.33	1.70	0.30
91	500.0	1.85	0.40	0.18	1.95	0.18
92	54.0	1.37	0.46	0.20	1.65	0.18
93	60.0	1.39	1.00	0.30	1.30	0.25
94	11.0	1.02	4.00	0.48	1.62	0.22
95					0.80	0.26
96					0.36	0.35
97					1.51	0.33
98					1.40	0.34
99					1.50	0.29
100					1.50	0.29
101					2.11	0.26
102					1.48	0.20
103					1.59	0.12
104					0.64	0.41
105					1.18	0.15
106					1.06	0.19
107					1.85	0.18
108					1.25	0.33
109					1.85	0.18
110					1.37	0.20
111					1.39	0.30
112					1.02	0.48
113	------	------	------	------	------	------

Worksheet C7C Probability calculations

```
 1  15 Sep 87, file C7C.WK1 on disks GSK 059-061
 2
 3  Worksheet C7C.WK1.   Probability calculations
 4
 5
 6  Part 1.   Abbreviated running instructions (Details in text)
 7
 8  (1)   To calculate the probability of gambler's ruin through
 9        a run of bad luck, press Alt-a.
10
11         Number of prospects to be explored:              10
12          Percent probability of success in
13               each exploration:                          20
14         Probability of ruin:                             11%
15
16  (2)   To calculate a factorial, press Alt-B.
17
18         Factorial number, N                               5
19         Factorial value, N!                             120
20
21
22
23
24
25
26
27
28
29
```

Worksheet C7C *(continued)*

```
30
31
32
33
34
35   Part 2.   List of macros
36
37   \a        {GOTO}C8~
38             {DOWN 3}~
39             {RIGHT 5}~
40              {GETNUMBER "Enter the number of prospects to be explor
41             {DOWN}
42              {GETNUMBER "Enter the percent probability of success i
43             {DOWN}
44              {LET RUIN,(1-(P/100))^N}~
45
46
47
48
49
50
51
52
53
54   Factorial macro (Ewing and others, 1987)
55
56   \b        {GOTO}C16~
57             {DOWN 2}~
58             {RIGHT 5}~
```

Worksheet C7C *(continued)*

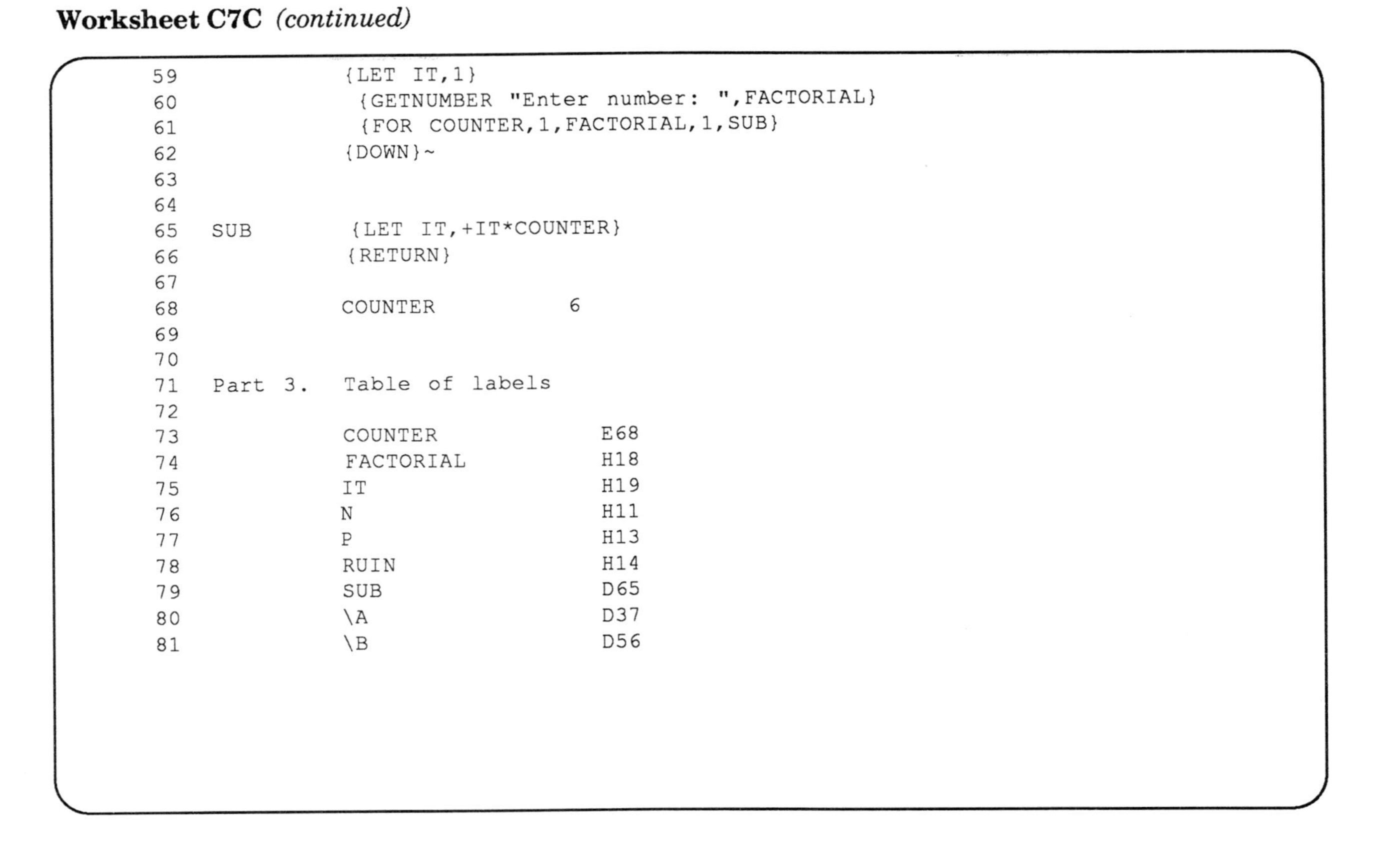

```
59              {LET IT,1}
60               {GETNUMBER "Enter number: ",FACTORIAL}
61               {FOR COUNTER,1,FACTORIAL,1,SUB}
62              {DOWN}~
63
64
65   SUB         {LET IT,+IT*COUNTER}
66              {RETURN}
67
68              COUNTER           6
69
70
71   Part 3.    Table of labels
72
73              COUNTER               E68
74              FACTORIAL             H18
75              IT                    H19
76              N                     H11
77              P                     H13
78              RUIN                  H14
79              SUB                   D65
80              \A                    D37
81              \B                    D56
```

CHAPTER 8

Compound Interest

In this chapter, I will first introduce compound-interest formulae, then provide a worksheet to tabulate discrete compounding interest factors at any selected rate of interest, and finally solve typical problems using these factors. The presentation and notation follows that of Frank Stermole of the Colorado School of Mines, who taught me economic evaluation and investment decision methods in a short course that "has been presented more than 300 times since 1970 to more than 10,000 practicing engineers, scientists, accountants, and managers working in petroleum, mineral, and general industry." (Quotation from course brochure.)

8.1 COMPOUND INTEREST FORMULAE

Stermole and Stermole (1987, p. 17) designate the factors according to this rule: the first letter in each factor designates the quantity that is calculated, whereas the second letter designates the quantity that is given. The subscripts are the period interest rate, I, followed by the number of interest compounding periods, N.

Table 8.1 Compound-interest factors and formulae. After Stermole and Stermole (1987, p. 22) with notation adjusted to 1-2-3's style.

Name	Notation	Formula
Single Payment Compound Amount Factor	F/P(I,N)	(1+I)^N
Single Payment Present-Worth Factor	P/F(I,N)	1/((1+I)^N)
Uniform Series Compound Amount Factor	F/A(I,N)	(((1+I)^N)-1)/I
Sinking-Fund Deposit Factor	A/F(I,N)	I/(((1+I)^N)-1)
Capital-Recovery Factor	A/P(I,N)	(I*((1+I)^N))/(((1+1)^N)-1)
Uniform Series Present-Worth Factor	P/A(I,N)	(((1+I)^N)-1)/ (I*((1+I)^N))

Notation (from Stermole and Stermole, 1987, p. 16)

- P: Present single sum of money
- F: A future single sum of money
- A: The amount of each payment in a uniform series of equal payments
- N: The number of interest compunding periods
- I: The period compound interest rate

All of these relationships are used in evaluation. Probably most familiar is the single payment compound-amount factor, F/P(I,N). We use this factor to calculate the future worth, F, after N years (or other period), of a present sum of money, P, with interest at I percent compounded per year (or other period). The formula is

$$F = P*(1 + i)^\wedge N.$$

We use the single payment present-worth factor to calculate the present value of a future sum, F, by solving the previous equation for P to obtain

$$P = F/((1 + I)^\wedge N).$$

We might use this formula to discount back to present time a series of payments to be received after a mine starts operation at some time in the future.

The uniform series compound amount factor, F/A(I,N), gives the future value of a series of equal payments. This formula will calculate the value at retirement of $2000-per-year IRA investments, assuming that interest rates are constant.

The sinking-fund deposit factor, A/F(I,N), calculates the series of payments required to provide a specified amount of money at some time in the future. For instance, you may wish to determine the payments necessary to accumulate the down-payment for a house at a specified time in the future.

The capital-recovery factor, A/P(I,N), gives the uniform series of payments equal to a present sum, for instance the house payments for a specified number of periods for a loan of a certain amount.

The uniform series present-worth factor, P/A(I,N) is used to determine the present single sum of money, P, that is equivalent to a uniform series of equal payments, A, for N periods at I percent interest per period. For our purposes, these periods are usually equal to years and in the rest of this book I use the terminology "years" instead of "periods." The formula is

$$P = A*((1+I)^N-1)/(I(1+I)^N)).$$

We could use this formula to determine the present value of a deposit that will produce a certain cash flow every year for a specified number of years.

The arithmetic gradient factor, not discussed in this book, is included to complete Stermole and Stermole's (1987) tables as explained in the next section.

8.2 TABLE OF DISCRETE COMPOUNDING FACTORS

Worksheet C8A.WK1 is a table of discrete compounding factors for from 1 to 30 years, set up in the format of Stermole and Stermole (1987, p. 431). In the worksheet, the interest rate is set at 10%, but you can change this in cell G8 to any other value. Of course, you may need to adjust the cell widths or formats in order to display your results.

Figures 8.1 to 8.3 graph the interest factors for values of N from 1 to 30.

8.3 SOLVING INTEREST PROBLEMS

1-2-3 can readily be programmed to solve interest problems. Worksheet C8B.WK1 provides the required formulae. Part 1 of this worksheet is a data-entry table. The values entered in cells E14 to E19 are processed in Part 2 to give the six discrete compounding factors in cells E29 to E41. 1-2-3 provides functions for the factors most used in business; the others are programmed. The examples are those in the file as stored on the floppy diskette for 10% interest and six years.

Part 3 of the worksheet solves ten typical problems discussed in the next paragraph. In the worksheet, the solution formulae are given in column E, and the values in this column are those obtained when the values in the data-entry table are those in the printed table (I = 0.1; N = 6; all other values = 1). The values in column F are those obtained when the values in Part 1 are set for the particular problem. They are copied from the values computed by formula in column E by using the 1-2-3 function /RV. For example, after setting the data-entry table for problem 8.1, the value 1.611 appears in cell E53. By following the prompts that appear when you key the command /RV, you can copy the value (not the formula) to cell F53.

Part 4 is the table of labels. Because of the way 1-2-3 alphabetizes, label 8.10 follows label 8.1 instead of label 8.9.

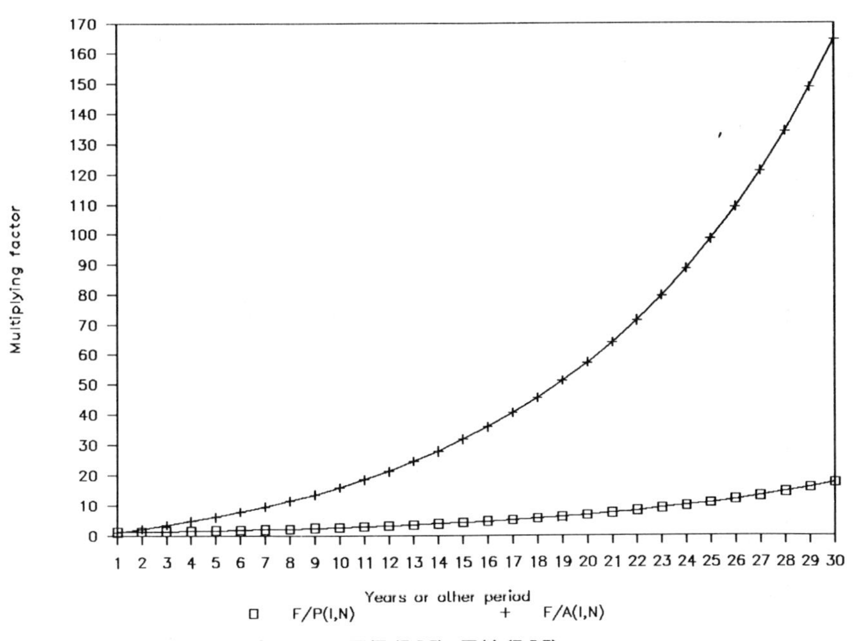

Figure 8.1 Interest factors, F/P(I,N), F/A(I,N)

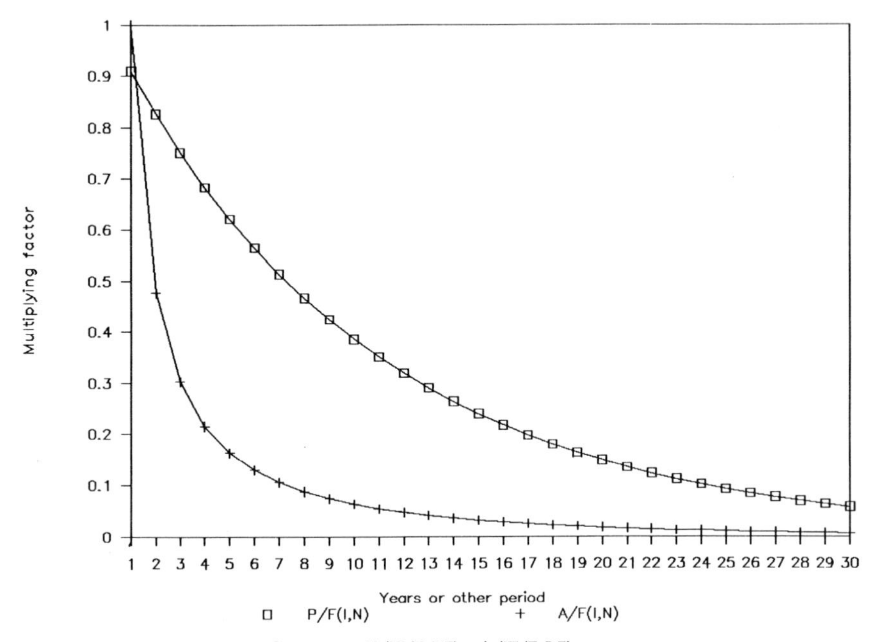

Figure 8.2 Interest factors, P/F(I,N), A/F(I,N)

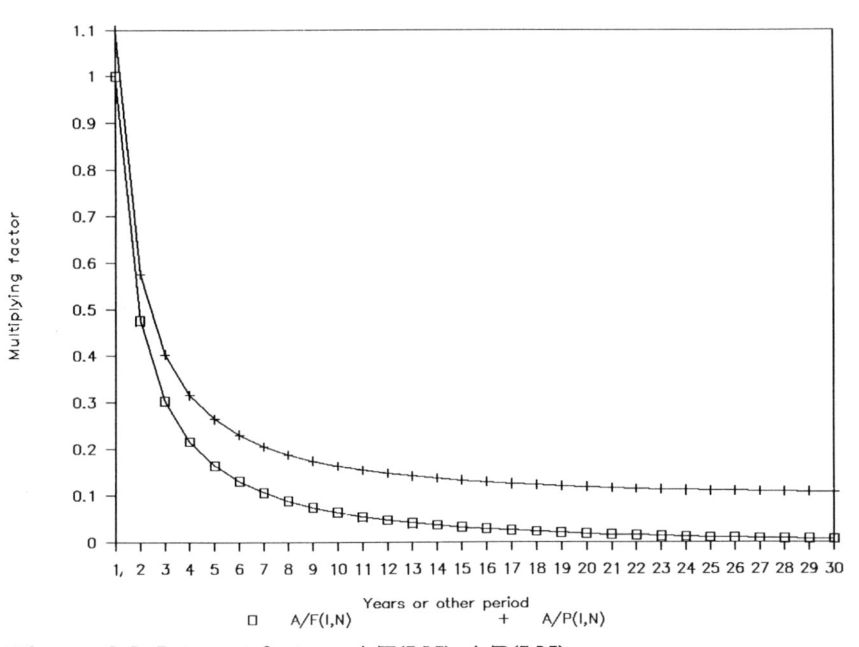

Figure 8.3 Interest factors, A/F(I,N), A/P(I,N)

If you practice with these problems, note that I have labeled the ranges and cells so that you can move around the worksheet using the GoTo function key (F5). For instance, if you press function key (F5) followed by 8.1, the cursor will move to problem 8.1.

The following is an explanation of the ten problems presented in Part 3:

8.1 Calculate the future value of $1. invested at 10% interest compounded annually for 5 years (Peters, 1987, p. 268). (Set E17 to 5.)

8.2 Suppose that you borrow $1 from your friendly loan shark with the agreement that you will repay him doubling the amount of the loan for every day you keep the money. That is, you will owe him $2 tomorrow, $4 the day after tomorrow, etc. Suppose further that although you fully intend to repay him on the day after tomorrow, you let the matter slide for 20 days. How much do you owe him? (Set E17 to 20 and E18 to 1.)

8.3 How much would you need to invest at 10 % interest compounded annually to yield $1 in five years (Peters, 1987, p. 269)? (Set E17 to 5 and E18 to 0.1.)

8.4 Calculate the uniform series of equal payments made at the end of each year for 15 years that are equivalent to a $10,000 payment 15 years from now if interest is 9 percent per year compounded annually. (Set E17 to 10000, E17 to 15, and E18 to 0.09.)

8.5 Calculate the present value of a series of $1000 payments to be made at the end of each year for 6 years if interest is 20 percent per year compounded annually. Repeat for 12 years. Note that doubling the number of years does not double the present value. (Set C16 to 1000, C17 to 6, and C18 to 0.2; then set C7 to 12.)

8.6 How much money (after taxes) do you need to win in the New York State Lottery in order to stop working for 20 years. Assume that you have no money now, that you will settle for an income of $30,000 a year, and that your winnings can be invested to yield 12% interest. Set E16 to 30000, E17 to 20, and E18 to 0.12.)

8.7 A mine is expected to pay dividends of $200,000 a year for ten years. A certain investor wishes 15% a year on his investment. What is the present value of the anticipated income at this rate of interest? (Set E16 to 200000, E17 to 10, and E18 to 0.15.)

8.8 A second investor wishes a 20% return on the investment of question 8.7. What is the present value in his situation? (Set E18 to 0.2.)

8.9 This problem is from Peters (1987, p. 270). "Take a capital investment with a starting date value of $4 million and consider that the investment is expected to return $2 million in the first and second years, $1 million in the third year, $0.5 million in the fourth and fifth years, and $1 million in the sixth year. The interest rate — hurdle rate — is 15% compounded annually. (Key the returns into range D76..D81, and calculate the discount factors in range E76. E81.)

8.10 Suppose that you, at age 30, decide to put $2000 a year into an Individual Retirement Account (IRA) to provide for your retirement at age 65. Suppose that you have two choices: (1) To start now, contribute $2000 a year for 8 years, and then stop, or, (2) To wait 8 years, and then contribute $2000 a year for the rest of your life until you retire? At 10% interest, which choice will give you a larger retirement fund? (In line 98, calculate the future value of your $2000 annual payments for 8 years; then, in line 99, calculate the future value of this amount (cell E98) after 27 years. In line 101, calculate the future value of $2000 annual payments for 27 years.)

Worksheet C8A Discrete compounding interest factors

```
 1   12 Jan 88, file C8A.WK1 on disks GSK 059-061
 2
 3   Worksheet C8A.WK1 for discrete compounding interest factors
 4
 5   Part 1.  Interest rate
 6
 7   Set the interest rate in cell
 8   G8 as a decimal fraction               0.1
 9
10
11    Part 2. Discrete compounding factors for ------>      10.0%
12
13    -----------------------------------------------------------------
14        F/P(I,N) P/F(I,N) F/A(I,N) A/F(I,N) A/P(I,N) P/A(I,N) A/G(I,N)
15    -----------------------------------------------------------------
16
17     1     1.100   0.9091     1.000  1.00000  1.10000    0.909
18     2     1.210   0.8264     2.100  0.47619  0.57619    1.736    0.476
19     3     1.331   0.7513     3.310  0.30211  0.40211    2.487    0.937
20     4     1.464   0.6830     4.641  0.21547  0.31547    3.170    1.381
21     5     1.611   0.6209     6.105  0.16380  0.26380    3.791    1.810
22     6     1.772   0.5645     7.716  0.12961  0.22961    4.355    2.224
23     7     1.949   0.5132     9.487  0.10541  0.20541    4.868    2.622
24     8     2.144   0.4665    11.436  0.08744  0.18744    5.335    3.004
25     9     2.358   0.4241    13.579  0.07364  0.17364    5.759    3.372
26    10     2.594   0.3855    15.937  0.06275  0.16275    6.145    3.725
27    11     2.853   0.3505    18.531  0.05396  0.15396    6.495    4.064
28    12     3.138   0.3186    21.384  0.04676  0.14676    6.814    4.388
29    13     3.452   0.2897    24.523  0.04078  0.14078    7.103    4.699
```

Worksheet C8A *(continued)*

```
30   14    3.797   0.2633   27.975  0.03575  0.13575    7.367     4.996
31   15    4.177   0.2394   31.772  0.03147  0.13147    7.606     5.279
32   16    4.595   0.2176   35.950  0.02782  0.12782    7.824     5.549
33   17    5.054   0.1978   40.545  0.02466  0.12466    8.022     5.807
34   18    5.560   0.1799   45.599  0.02193  0.12193    8.201     6.053
35   19    6.116   0.1635   51.159  0.01955  0.11955    8.365     6.286
36   20    6.727   0.1486   57.275  0.01746  0.11746    8.514     6.508
37   21    7.400   0.1351   64.002  0.01562  0.11562    8.649     6.719
38   22    8.140   0.1228   71.403  0.01401  0.11401    8.772     6.919
39   23    8.954   0.1117   79.543  0.01257  0.11257    8.883     7.108
40   24    9.850   0.1015   88.497  0.01130  0.11130    8.985     7.288
41   25   10.835   0.0923   98.347  0.01017  0.11017    9.077     7.458
42   26   11.918   0.0839  109.182  0.00916  0.10916    9.161     7.619
43   27   13.110   0.0763  121.100  0.00826  0.10826    9.237     7.770
44   28   14.421   0.0693  134.210  0.00745  0.10745    9.307     7.914
45   29   15.863   0.0630  148.631  0.00673  0.10673    9.370     8.049
46   30   17.449   0.0573  164.494  0.00608  0.10608    9.427     8.176
47   --------------------------------------------------------------------
48
49  Part 3.   Table of range names
50
51  #A.F.I.N      G17
52  #A.G.I.N      J18
53  #A.P.I.N      H17
54  #F.A.I.N      F17
55  #F.P.I.N      D17
56  #P.A.I.N      I17
57  #P.F.I.N      D17
58  I             G8
```

Worksheet C8B Solving interest problems

```
 1  29 May 89, file C8B.WK1 on disks GSK 059-061
 2
 3  Worksheet C8B.WK1. Solving interest problems
 4
 5  Part 1.  Data-entry table.
 6      Enter the required values in column E.
 7
 8     C                   D                    E
 9  ---------------------------------------------------
10    Range            Description            Value
11     Name
12  ---------------------------------------------------
13
14      P             Present value                1
15      F              Future value                1
16      A                Annuity                   1
17      N        Number of periods (years)         6
18      I             Interest rate              0.1
19      G        Arithmetic gradient series        1
20  ---------------------------------------------------
21
22
```

Worksheet C8B *(continued)*

```
23   Part 2.  Discrete compounding factors.
24
25    -----------------------------------------------------------
26     Factor                  Formula             Example (I=0.1, N=6)
27    -----------------------------------------------------------
28
29    F/P(I,N)   (P)*((1+I)^N)                            1.772
30
31    P/F(I,N)   (F)*(1/(1+I)^N)                          0.5645
32
33    F/A(I,N)   @FV(A,I,N)                               7.716
34
35    A/F(I,N)   (F)*((I)/(((1+I)^N)-1))                  0.12961
36
37    A/P(I,N)   @PMT(P,I,N)                              0.22961
38
39    P/A(I,N)   @PV(A,I,N)                               4.355
40
41    A/G(I,N)   (G)*((1/I)-(N/I)*(#A.F.I.N))             10.000
42    -----------------------------------------------------------
43
44
```

Worksheet C8B *(continued)*

```
45   Part 3.   Solutions to problems
46
47      C                     D                      E          F
48   ---------------------------------------------------------------
49    Problem                Factor              Solution   Solution
50     Number                                     Formula     Value
51   ---------------------------------------------------------------
52
53   8.1          (P)*((1+I)^N)                    $1.772      $1.611
54
55   8.2          (P)*((1+I)^N)                        $2  $1,048,576
56
57   8.3          (F)*(1/(1+I)^N)                 $0.5645     $0.6209
58
59   8.4          (F)*((I)/(((1+I)^N)-1))           $0.13     $340.59
60
61   8.5          @PV(A,I,N)                           $4      $3,326
62                                                     $4      $4,439
63
64   8.6          @PV(A,I,N)                           $4    $224,083
65
66   8.7          @PV(A,I,N)                           $4  $1,003,754
67
68   8.8          @PV(A,I,N)                           $4    $838,494
69
```

Worksheet C8B *(continued)*

```
70  8.9
71
72   ---------------------------------------------------------------------
73       Year                  Income,              15% Discount  Present
74                               $M                    Factor       Value
75   ---------------------------------------------------------------------
76             1                            2.0          0.870  $1,739,130
77             2                            2.0          0.756  $1,512,287
78             3                            1.0          0.658    $657,516
79             4                            0.5          0.572    $285,877
80             5                            0.5          0.497    $248,588
81             6                            1.0          0.432    $432,328
82                                                               ----------
83      Totals                              7.0                 $4,875,727
84  Less capital investment                                     ($4,000,000)
85                                                               ----------
86  Net present value (acquisition value)                         $875,727
87
88   ---------------------------------------------------------------------
89
90
```

Worksheet C8B *(continued)*

```
 91  8.10
 92
 93   ---------------------------------------------------------
 94  Age                                                     30
 95  Years until retirement                                  35
 96  Years until retirement - 8                              27
 97
 98               @FV(2000,0.1,8)                       $22,872
 99                (E98)*((1.1)^E96)                   $299,849
100
101                 @FV(2000,0.1,E96)                  $242,200
102    ---------------------------------------------------------
103
104
105   Part 4.   Table of labels
106
107   #A.F.I.N     C55
108   #A.P.I.N     E37
109   #F.A.I.N     E33
110   #F.P.I.N     E29
111   #P.A.I.N     E39
112   #P.F.I.N     E31
113   8.1          C53
114   8.10         C91
115   8.2          C55
116   8.3          C57
117   8.4          C59
118   8.5          C61
119   8.6          C64
```

Worksheet C8B *(continued)*

```
120  8.7      C66
121  8.8      C68
122  8.9      C70
123  A        E16
124  F        E15
125  G        E19
126  I        E18
127  N        E17
128  P        E14
129  PART.1   C14..E19
130  PART.2   C29..E41
131  PART.3   C45
132  PART.4   C93
```

CHAPTER 9

Depreciation and Depletion

In the United States, "depreciation" and "depletion" are accounting terms used to calculate income tax. They are needed to determine cash flow, which is the basis for any economic evaluation.

Depreciation, as used in economic evaluation, "is a tax allowance that is a reasonable allowance for the exhaustion, wear and tear and obsolescence of property used in a trade or business, or of property held by a tax payer for the production of income" (Stermole and Stermole, 1987, p. 229).

Depletion is the "recovery of an owner's economic interest in mineral (including oil and gas) reserves through federal tax deductions related to removal of the mineral over the economic life of the property" (Stermole and Stermole, 1987, p. 455).

Neither depreciation or depletion involve a corresponding flow of money from the enterprise, any more than personal income tax deductions for individuals, blindness, age, etc. exactly represent corresponding expenditures of money.

Depreciation and depletion are explained briefly by Peters (1987, p. 279-280) and in detail by Stermole and Stermole (1987, p.222-256).

9.1 YEARLY NET INCOME AND CASH FLOW CALCULATION

Worksheet C9A.WK1 is a yearly income and cash-flow calculation based on the format of Peters (1987, p. 279), with additional lines to calculate cost depletion. Part 1 gives fixed data for the project; row 11 assumes a property cost of $1 million, which is used to calculate cost depletion.

Part 2 includes all of Peters's (1987, p. 279) entries, as well as additional ones to calculate cost depletion. In line 26, depreciation is treated as straightline, the method usually used in preliminary work. Stermole and Stermole (1987, p. 229-241) explain several alternative methods, which may be necessary for detailed work. 1-2-3 provides special functions for the *straightline*, *double-declining-balance*, and *sum-of-the-years'-digits* methods.

Depletion is calculated in rows 30 to 36. Allowable percentage depletion is calculated in rows 30 to 32. Cell D30 calculates percentage depletion by multiplying the percentage factor from D14 by the revenue in D22, and cell D31 calculates the 50% allowable limit. Cell D32 contains the smaller of the values in the two previous cells. Cost depletion is calculated in cell D34. Cell D36 contains the allowable depletion, which is the larger of the values in cells D32 and D34.

Subtracting selected depletion (D36) from net income before depletion (D28) gives taxable income (D38), and subtracting income tax (D39) gives net income in cell D41.

Finally, we obtain cash flow (D45) by adding back depreciation and depletion in rows 42 and 43, remembering that these two items were subtracted only to calculate income tax and do not in themselves represent actual expenditures of cash.

9.2 THE RIDGEWAY, SOUTH CAROLINA, GOLD MINE

A note in Engineering and Mining Journal (E&MJ) for May, 1987 (p. 78) states that "A feasibility report issued by the Ridgeway Mining Co. to Amselco Minerals Inc. and Galactic Resources Ltd. projects commercial gold production at the Ridgeway, South Carolina, project starting by mid-1988 at a rate of 158,000 tr oz/yr of gold over the first four years (1989-92), and 133,000 tr oz/yr over the 11-year life of the mine. Amselco and Galactic are joint venture partners in the project, 15 miles northeast of Columbia. Capital costs to bring the Ridgeway project into production are estimated at $76 million."

Using these data (supplemented by information from company reports), we can construct worksheet C9B.WK1. To the items in Part 1 of worksheet C9A.WK1, we add production, selling price, and operating cost. The worksheet formulation of this problem illustrates the flexibility of the approach; the assumptions can readily be changed — now by knowledgeable people — or later as more data are published.

Worksheet C9A Yearly net income and cash flow

```
 1  12 Jan 88, file C9A.WK1 on disks GSK 059-61
 2
 3  Worksheet C9A.WK1.  Yearly net income and cash flow
 4       (after Peters, 1987, p. 279)
 5
 6
 7  Part 1.  Fixed data
 8
 9               C                         D
10   ------------------------------------------
11  Property cost                      $1,000,000
12  Capital investment                 $8,500,000
13  Years life                                 10
14  Depletion percentage                       15
15  Income tax percentage                      55
16   ------------------------------------------
17
18
```

Worksheet C9A *(continued)*

```
19  Part 2.  Yearly net income and cash flow
20
21  ---------------------------------------------
22  Revenue                            $15,420,000
23  Operating costs                      7,340,000
24
25  Operating income                     8,080,000
26  Depreciation                           850,000
27
28  Net income before depletion          7,230,000
29
30  Percentage depletion                 2,313,000
31  50% limit                            3,615,000
32  Selected percentage depletion        2,313,000
33
34  Cost depletion                         100,000
35
36  Selected depletion                   2,313,000
37
38  Taxable income                       4,917,000
39  Income tax @ 55 %                    2,704,350
40
41  Net income                           2,212,650
42  Depreciation                           850,000
43  Selected depletion                   2,313,000
44
45  Cash flow                            5,375,650
46  ---------------------------------------------
47
```

Worksheet C9A *(continued)*

```
48
49   Part 3.  Table of labels
50
51   !CASH.FLOW                      D45
52   !COST.DPL                       D34
53   !DEPRE.&.AMORT.                 D26
54   !DPL.%.SELECTED                 D32
55   !DPL.50%.LIMIT                  D31
56   !DPL.SELECTED                   D36
57   !INCOME.TAX                     D39
58   !NET.BEFORE.DPL                 D28
59   !NET.INCOME                     D41
60   !OP.INCOME                      D25
61   !OPERATING.COST                 D23
62   !PCT.DPL                        D30
63   !REVENUE                        D22
64   !TAX.INCOME                     D38
65   %DEPLETION                      D14
66   %INCOME.TAX                     D15
67   CAP.INVESTMENT                  D12
68   LIFE                            D13
69   PROPERTY.COST                   D11
```

Worksheet C9B Yearly net income and cash flow. Ridgeway, South Carolina, gold mine

```
 1   19 Feb 88, file C9B.WK1 on disks GSK 059-061
 2
 3   Worksheet C9B.WK1.  Yearly net income and cash flow,
 4       Ridgeway, South Carolina, gold mine
 5
 6
 7   Part 1.  Fixed data
 8
 9               C                              D
10    ----------------------------------------------
11   Property cost                        $18,000,000
12   Capital investment                   $76,000,000
13   Years life                                    12
14   Depletion percentage                          15
15   Income tax percentage                         55
16   Production, tr oz/year                   135,000
17   Selling price, $/tr oz                       420
18   Operating costs, $/tr oz                     210
19
20
21
22
23    ----------------------------------------------
24
25
```

Worksheet C9B *(continued)*

```
26  Part 2.  Yearly net income and cash flow
27
28  ---------------------------------------------
29  Revenue                          $56,700,000
30  Operating costs                   28,350,000
31
32  Operating income                  28,350,000
33  Depreciation                       6,333,333
34
35  Net income before depletion       22,016,667
36
37  Percentage depletion               8,505,000
38  50% limit                         11,008,333
39  Selected percentage depletion      8,505,000
40
41  Cost depletion                     1,500,000
42
43  Selected depletion                 8,505,000
44
45  Taxable income                    13,511,667
46  Income tax @ 55 %                  7,431,417
47
48  Net income                         6,080,250
49  Depreciation                       6,333,333
50  Selected depletion                 8,505,000
51
52  Cash flow                         20,918,583
53  ---------------------------------------------
54
```

Worksheet C9B *(continued)*

```
55
56  Part 3. Table of labels
57
58  !CASH.FLOW                      D52
59  !COST.DPL                       D41
60  !DEPRE.&.AMORT.                 D33
61  !DPL.%.SELECTED                 D39
62  !DPL.50%.LIMIT                  D38
63  !DPL.SELECTED                   D43
64  !INCOME.TAX                     D46
65  !NET.BEFORE.DPL                 D35
66  !NET.INCOME                     D48
67  !OP.INCOME                      D32
68  !OPERATING.COST                 D30
69  !PCT.DPL                        D37
70  !REVENUE                        D29
71  !TAX.INCOME                     D45
72  %DEPLETION                      D14
73  %INCOME.TAX                     D15
74  CAP.INVESTMENT                  D12
75  LIFE                            D13
76  OP.COST.PER.OZ                  D18
77  PRICE.PER.OZ                    D17
78  PRODUCTION                      D16
79  PROPERTY.COST                   D11
```

CHAPTER 10

Discounted Cash Flow Rate of Return

The Discounted cash flow rate of return (DCFROR), also named the discounted cash flow return on investment (DCFROI), or the internal rate of return (IRR), is the method most widely used to compare investment opportunities. We may want to compare one mining venture with another or a mining venture with some other investment.

The discounted cash flow rate of return is simply the interest rate, I, in the formula

P = A(1)/(1+I)^1 + A(2)/(1+I)^2 + ... + A(N)/(1+I)^N

where P is the principal (capital investment), A(N) is the yearly cash flow (which may be positive, zero, or negative), I is the interest rate, and N is the number of years.

1-2-3 makes calculation of DCFROR easy by providing a function, which is explained in the next section. Peters (1987, p. 276-279) gives a brief account of DCFROR; Stermole and Stermole (1987, p. 257-283) provide a chapter that thoroughly explains the principle and details; Koch and Link (1971, p. 312-314) work a simple example in detail.

10.1 A SIMPLE EXAMPLE

Worksheet C10A.WK1 solves a simple example problem to illustrate the method. Assume that a truck with an expected useful life of 4 years costs $10,000. Annual income each year of the life of this truck is expected to be $8,000, and annual operating costs are expected to be $3,500. Straightline depreciation is used, and the overall tax rate is 50 percent.

Part 1 of the worksheet lists the fixed data for the problem; part 2 gives the yearly net income and cash flow. Part 3 is the calculation of DCFROR through 1-2-3's function

@IRR(estimate,range).

The range, D39..D43, is labeled CASH.FLOWS.0-4; the estimate is required, because 1-2-3 calculates the value of the function interactively. After 20 iterations, the entry ERR appears if a result to within 0.0000001 is not located. If, while using the @IRR function, you get the message ERR, simply substitute another, closer estimate of DCFROR. For precautions, see a 1-2-3 reference manual. From my initial estimate of 30%, recorded in the function as 0.3, 1-2-3 calculates the correct value of 15. This is exactly the value that makes the net present value equal to 0 at a hurdle rate of 15%.

10.2 A SECOND EXAMPLE OF DCFROR

A second example of DCFROR is provided in worksheet C10B.WK1 which presents a problem from Peters (1987, p. 278). Except for roundoff errors, the values in the worksheet match those of Peters.

In Part 1, we first determined net present value (NPV) at two interest rates that bracket DCFROR. For the data given, these rates are 16% and 17%.

DCFROR is calculated in Part 2, first by linear interpolation, and then by 1-2-3's formula. The results are carried out to several decimal places in order to demonstrate that linear interpolation does not give a precisely correct result, although it is certainly close enough for a preliminary evaluation. For one thing, none of the cash flows will be known exactly! Of course, calculation by 1-2-3's formula is faster, particularly if you cannot guess the interest rates that bracket DCFROR.

The worksheet is a convenient one to use to gain an understanding of DCFROR by changing the flows in the range D14..D21, recalculating, and seeing how DCFROR changes.

Worksheet C10A Illustration of DCFROR

```
 1   12 Jan 88, file C10A.WK1 on disks GSK 059-061
 2
 3   Worksheet C10A.WK1.   Illustration of DCFROR
 4
 5
 6   Part 1.   Fixed data
 7
 8                 C                          D
 9    --------------------------------------------
10    Cost of the truck                    $10,000
11    Life, years                                4
12    Income tax rate, %                        50
13    Estimate of DCFROR, %                     20
14     --------------------------------------------
15
16
```

Worksheet C10A *(continued)*

```
17   Part 2.  Yearly net income and cash flow
18
19   ------------------------------------------
20   Revenue                            $8,000
21   Operating costs                     3,500
22
23   Operating income                    4,500
24   Depreciation                        2,500
25
26   Taxable income                      2,000
27   Income tax                           1000
28
29   Net income                          1,000
30   Depreciation                        2,500
31
32   Cash flow                          $3,500
33   ------------------------------------------
34
35
36   Part 3.  Calculation of DCFROR
37
38   ------------------------------------------
39   Cash flow (Year 0)               (10,000)
40   Cash flow (Year 1)                  3,500
41   Cash flow (Year 2)                  3,500
42   Cash flow (Year 3)                  3,500
43   Cash flow (Year 4)                  3,500
44
45   DCFROR                                15%
```

Worksheet C10A *(continued)*

```
46   -------------------------------------------
47
48
49   Part 4.   Table of labels
50
51   %EST.DCFROR                    D13
52   CASH.FLOW                      D32
53   CASH.FLOW.0                    D39
54   CASH.FLOW.1                    D40
55   CASH.FLOW.2                    D41
56   CASH.FLOW.3                    D42
57   CASH.FLOW.4                    D43
58   CASH.FLOWS.0-4                 D39..D43
59   DEPRECIATION                   D24
60   INCOME.TAX                     D27
61   LIFE                           D11
62   NET.INCOME                     D29
63   OP.COST                        D21
64   OP.INCOME                      D23
65   REVENUE                        D20
66   TAXABLE.INCOME                 D26
67   TRUCK.COST                     D10
```

Worksheet C10B Estimating DCFROR

```
 1   12 Jan 88, file C10B.WK1 on disks GSK 059-061
 2
 3   Worksheet C10B.WK1.  Estimating DCFROR
 4       Problem from Peters (1987, p. 278)
 5
 6
 7   Part 1.  Calculation of present value at 16% and 17%
 8
 9            C       D       E       F       G       H
10    ------------------------------------------------------------------------
11     Year   Cash Flow    P/F(16,N)    Present       P/F(17,N)     Present
12                                      Value,16%                   Value,17%
13    ------------------------------------------------------------------------
14     0      $-230,000
15     1       50,000       0.862  $43,103        0.855  $42,735
16     2       70,000       0.743  52,021         0.731  51,136
17     3       70,000       0.641  44,846         0.624  43,706
18     4       60,000       0.552  33,137         0.534  32,019
19     5       50,000       0.476  23,806         0.456  22,806
20     6       40,000       0.410  16,418         0.390  15,594
21     7       60,000       0.354  21,230         0.333  19,992
22    ------------------------------------------------------------------------
23
24     Total present value         $234,561              $227,987
25     Capital investment          -230,000              -230,000
26
27     Net present value           $4,561                $-2,013
28    ------------------------------------------------------------------------
```

Worksheet C10B *(continued)*

```
29
30
31     Part 2.  Calculation of DCFROR
32
33    ------------------------------------------------------------------------
34     a.   Calculation by linear interpolation
35
36     Interpolation               16%+(4,561/(4,561+2,013))
37
38     Discounted cash flow rate of return (DCFROR) =              16.743%
39    ------------------------------------------------------------------------
40
41     b.   Calculation by 1-2-3's function @IRR
42
43                           @IRR(.3,D14..D21)                     16.738%
44    ------------------------------------------------------------------------
45
46
47     Part 3.  Table of labels
48
49     CASH.FLOW.0          D14
50     CASH.FLOW.1          D15
51     CASH.FLOW.2          D16
52     CASH.FLOW.3          D17
53     CASH.FLOW.4          D18
54     CASH.FLOW.5          D19
55     CASH.FLOW.6          D20
56     CASH.FLOW.7          D21
57     NPV.16               F27
```

Worksheet C10B *(continued)*

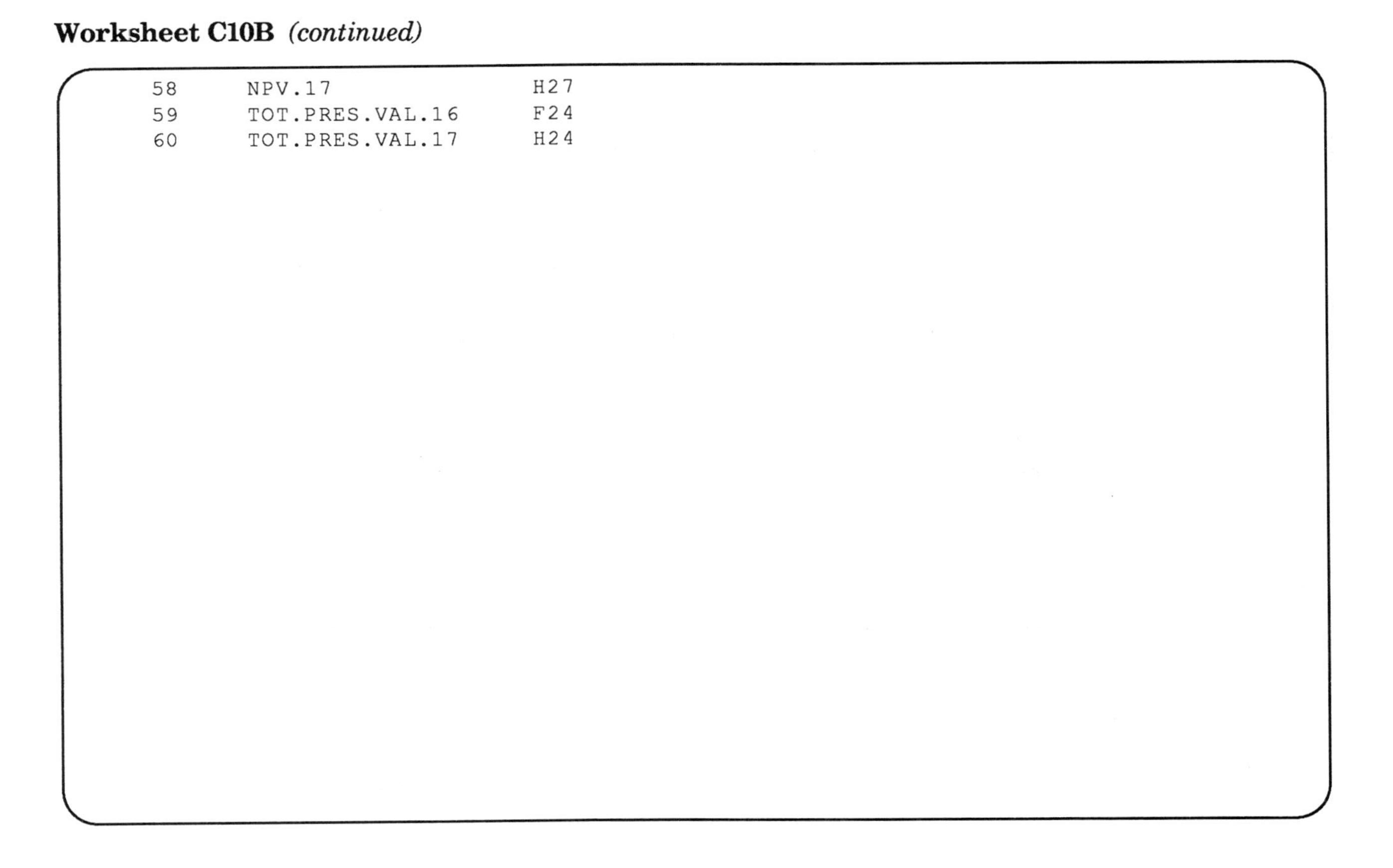

58	NPV.17	H27
59	TOT.PRES.VAL.16	F24
60	TOT.PRES.VAL.17	H24

CHAPTER 11

Blocking Ore From Data in Development Workings

In this chapter, we will first calculate grade and tonnage of ore from data obtained in development workings, and then calculate specific gravity of ore from its mineralogical composition.

11. 1 GRADE AND TONNAGE OF ORE FROM DEVELOPMENT DATA

The following sample averages represent four blocks of silver ore in the Carlos Francisco Mine at Casapalca, Peru. The development workings in this part of the mine consist of three levels, numbered from the top downward, 6, 7, and 8; and three raises, lettered from north to south, A, B, and C. Table 11.1 gives the length and width dimensions of these workings, together with the silver assays.

Table 11.1 Development data for 4 blocks of ore in the Carlos Francisco Mine.

Working		Length, m	Width, m	Silver, g/T
Level 6	Raise A to Raise B	110	0.9	300
Level 6	Raise B to Raise C	75	1.1	480
Level 7	Raise A to Raise B	110	1.0	540
Level 7	Raise B to Raise C	75	1.1	660
Level 8	Raise A to Raise B	110	1.0	750
Level 8	Raise B to Raise C	75	1.2	600
Raise A Level 6-7		40	0.9	240
Raise B Level 6-7		40	1.5	600
Raise C Level 6-7		40	1.2	540
Raise A Level 7-8		35	1.0	690
Raise B Level 7-8		35	1.2	660
Raise C Level 7-8		35	1.2	720

Sketching a longitudinal section of the four blocks will show you the geological situation. Additional information is this: (1) assays already have been adjusted to the minimum mining width, 0.9 m; (2) the specific gravity is 3.2; (3) the cut-off grade of ore is 450 grams silver per tonne.

From these data, we will calculate:

1. Tonnage of ore and its grade in each of the four blocks.

2. Aggregate tonnage and average grade of the four blocks.

3. Aggregate tonnage and grade after allowing for 10 percent dilution of barren wall rock.

Peters (1987, p. 482-491) explains the basic procedure, which is a weighted average. Using 1-2-3, we can set up the calculations (worksheet C11A.WK1). Part 1 gives the calculations for the block between levels 6 and 7 and raises A and B. Column C lists the arbitrary numbers designating the four sides of block 1; column D the lengths, column E the widths, and column F the assays. Column G lists the products of lengths times widths, and column H the products of lengths times widths times assays. The column sums, in line 17, then are used to calculate the estimates of assay and tonnes in columns I and J.

Because the grade for this block is lower than the cutoff grade, we need to reduce its size as shown in Part 2. The conventional procedure of defining a triangular block is followed.

Once we have set up the worksheet for one block, we can copy the format for the remaining four blocks using the /COPY command.

Part 6 summarizes the calculations for the four blocks. The sums in row 94 are used to calculate the average assay of 612 grams per tonne in row 97. Line 98 gives the 10 percent additional tonnage resulting from dilution of wall rock, which is assumed to contain no silver. Finally, row 100 gives the mined grade.

The table of labels, Part 7, applies only to Block 1.

11.2 CALCULATION OF SPECIFIC GRAVITY

Starting with the ore mineralogy and the grade expressed as a percentage of metal, we can calculate the specific gravity of the ore. The method is explained by Peters (1987, p. 481) and by Koch and Link, (1971, p. 249); the method in worksheet C11B.WK1 is that of the second reference.

Although the calculations are not complicated, formulating the problem seems to puzzle many people. I think it is helpful to present the problem in the concrete terms of determining the number of grams of each mineral in 100 grams of ore (rows 16 and 17 of worksheet C11B.WK1.)

Once the specific gravity is determined (row 10), you can determine other factors, such as the number of cubic feet per short ton, if required, using tables in Peters (1987) or other standard books.

The function relating grade and specific gravity is not linear. Figure 11.1 graphs the quartz/galena curve for various grades of galena.

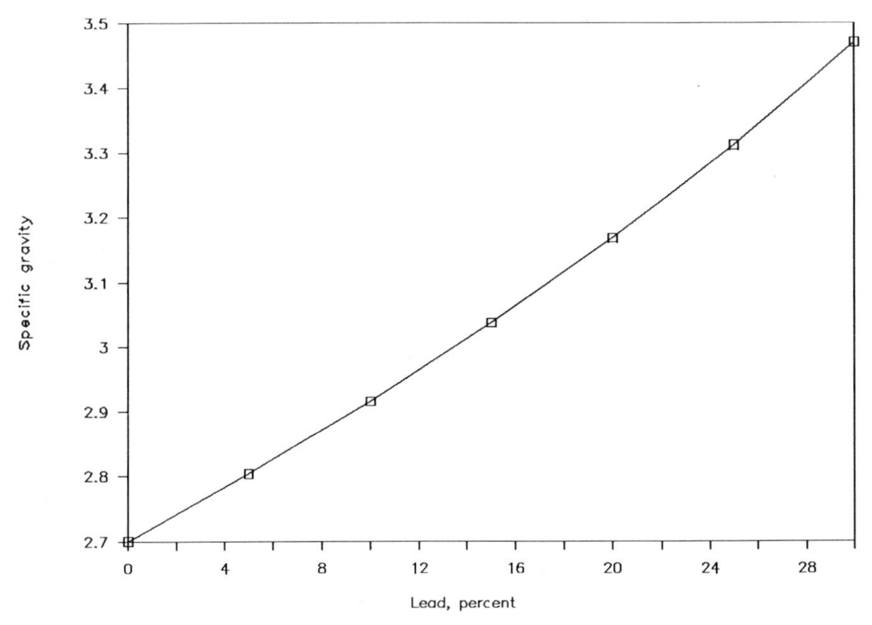

Figure 11.1 S.G. - grade relationship, example of a lead/quartz ore

Worksheet C11A Calculating grade and tonnage of ore

```
 1  02 Feb 89, file C11A.WK1 on disks GSK 059-061
 2
 3  Worksheet C11.A.WK1.  Calculating grade and tonnage of ore
 4
 5  Part 1.  Calculations for block 1 (Levels 6 and 7, Raises A and B)
 6
 7   C      D        E        F        G         H          I        J
 8  --------------------------------------------------------------------
 9  Side  Length,  Width,   Assay,    L x W   L x W x A    Block    Block
10   No.     m        m       g/T                          assay   tonnes
11  --------------------------------------------------------------------
12    1     110      0.9      300      99.0     29,700
13    2      40      1.5      600      60.0     36,000
14    3     110      1.0      540     110.0     59,400
15    4      40      0.9      240      36.0      8,640
16
17        300.0      1.0              305.0    133,740      438   14,315
18  --------------------------------------------------------------------
19
20
```

Worksheet C11A *(continued)*

```
21  Part 2.  Size of block 1 reduced to raise grade above cutoff 22
23  -------------------------------------------------------------------
24  Side  Length,   Width,    Assay,     L x W   L x W x A   Block    Block
25  No.      m         m       g/T                            assay   tonnes
26  -------------------------------------------------------------------
27    2       40       1.5      600       60.0     36,000
28    3      110       1.0      540      110.0     59,400
29
30         150.0       1.1   1140.0      170.0     95,400      561     7,979
31  -------------------------------------------------------------------
32
33
34  Part 3.  Calculations for block 2 (Levels 6 and 7, Raises B and C)
35
36   C        D         E         F         G         H          I        J
37  -------------------------------------------------------------------
38  Side  Length,   Width,    Assay,     L x W   L x W x A   Block    Block
39  No.      m         m       g/T                            assay   tonnes
40  -------------------------------------------------------------------
41    1       75       1.1      480       82.5     39,600
42    2       40       1.2      540       48.0     25,920
43    3       75       1.1      660       82.5     54,450
44    4       40       1.5      600       60.0     36,000
45
46         230.0       1.2               273.0    155,970      571    11,395
47  -------------------------------------------------------------------
48
49
```

Worksheet C11A *(continued)*

```
50  Part 4.  Calculations for block 3 (Levels 7 and 8, Raises B and C)
51
52   C       D        E         F        G         H          I        J
53  ------------------------------------------------------------------------
54  Side  Length,  Width,    Assay,    L x W   L x W x A    Block    Block
55   No.     m        m        g/T                          assay   tonnes
56  ------------------------------------------------------------------------
57    1       75      1.1       660     82.5      54,450
58    2       35      1.2       720     42.0      30,240
59    3       75      1.2       600     90.0      54,000
60    4       35      1.2       660     42.0      27,720
61
62         220.0      1.2              256.5     166,410      649    9,794
63  ------------------------------------------------------------------------
64
65
66  Part 5.  Calculations for block 4 (Levels 7 and 8, Raises A and B)
67
68   C       D        E         F        G         H          I        J
69  ------------------------------------------------------------------------
70  Side  Length,  Width,    Assay,    L x W   L x W x A    Block    Block
71   No.     m        m        g/T                          assay   tonnes
72  ------------------------------------------------------------------------
73    1      110      1.0       540    110.0      59,400
74    2       35      1.2       660     42.0      27,720
75    3      110      1.0       750    110.0      82,500
76    4       35      1.0       690     35.0      24,150
77
78         290.0      1.0              297.0     193,770      652   12,617
```

Worksheet C11A *(continued)*

```
 79   ------------------------------------------------------------------
 80
 81
 82  Part 6.   Summary calculations for the 4 blocks
 83
 84   C       D        E        F        G         H          I        J
 85   ------------------------------------------------------------------
 86  Block                                       Tonnes     Block    Block
 87   No.                                       x Assay     assay   tonnes
 88   ------------------------------------------------------------------
 89    1                                      4,477,440      561     7,979
 90    2                                      6,510,052      571    11,395
 91    3                                      6,353,836      649     9,794
 92    4                                      8,231,884      652    12,617
 93
 94                                          25,573,213              41,784
 95   ------------------------------------------------------------------
 96
 97                Ore grade, g/T                            612
 98                10% dilution                                     4,178
 99                Tonnage after dilution                          45,963
100                 Mined grade, g/T                         556
101   ------------------------------------------------------------------
102
103
```

Worksheet C11A *(continued)*

```
104  Part 7.  Table of labels
105
106  AG.1                    F12
107  AG.2                    F27
108  AG.3                    F28
109  AG.4                    F15
110  BLOCK.L.X.W.X.A         H30
111  L.X.W.1                 G12
112  L.X.W.2                 G27
113  L.X.W.3                 G28
114  L.X.W.4                 G15
115  L.X.W.BLOCK             G30
116  L.X.W.X.A.1             H12
117  L.X.W.X.A.2             H27
118  L.X.W.X.A.3             H28
119  L.X.W.X.A.4             H15
120  LENGTH.1                D12
121  LENGTH.2                D27
122  LENGTH.3                D28
123  LENGTH.4                D15
124  LENGTH.BLOCK            D30
125  WIDTH.1                 E12
126  WIDTH.2                 E27
127  WIDTH.3                 E28
128  WIDTH.4                 E15
129  WIDTH.BLOCK             E30
```

Worksheet C11B Calculating specific gravity of ore

```
 1  18 Feb 88, file C11B on disks GSK 059-061
 2
 3  Worksheet C11B.WK1.  Calculating specific gravity of ore
 4
 5  Part 1.  Calculation of specific gravity for a galena,
 6      quartz ore (from Koch and Link, 1971, p. 249)
 7
 8   ----------------------------------------------------
 9  Grade of ore, %                                     5
10   Specific gravity of ore                       2.8036
11   ----------------------------------------------------
12   Atomic weight of Pb                            207.2
13   Atomic weight of sulfur                         32.1
14   Molecular weight galena                        239.3
15   Percent of Pb in galena                       86.586
16   Grams of galena in 100 grams of ore            5.775
17   Grams of quartz in 100 grams of ore           94.225
18   Specific gravity of galena                       7.5
19   Specific gravity of quartz                       2.7
20   Volume of galena                               0.770
21   Volume of quartz                              34.898
22   Volume of ore                                 35.668
23   ----------------------------------------------------
24
25
```

Worksheet C11B *(continued)*

```
26   Part 2.   Table of labels
27
28   GN.G.IN.ORE                                   D16
29   GN.MOL.WT                                     D14
30   GN.S.G.                                       D18
31   GN.VOLUME                                     D20
32   ORE.VOLUME                                    D22
33   PB.%                                          D9
34   PB.%.IN.GN                                    D15
35   PB.ATOMIC.WT                                  D12
36   QTZ.G.IN.ORE                                  D17
37   QTZ.S.G.                                      D19
38   QTZ.VOLUME                                    D21
39   S.ATOMIC.WT                                   D13
```

CHAPTER 12

Blocking Ore from Drillhole Data

This chapter provides worksheets to block ore from drillhole data by making linear and quadratic regressions. I will introduce the method in section 12.1 with a small, simple example of fitting a regression model to structural geologic data.

The example of blocking ore is based on a data set from the Chambishi copper mine in Zambia (Koch, 1975, 1978; Mendelsohn, 1980). Figure 12.1 is a vertical longitudinal section that shows the open pit, blocks of ore below the open pit, and the intersections of 38 diamond-drill holes with the ore-bearing sedimentary bed. Next to each intersection is plotted the grade in copper at that point. We will calculate a quadratic regression and solve the resulting equation for the grade of each block. Then, we will compare this estimate with estimates made in other ways.

12.1 QUADRATIC REGRESSION — RANGELY, COLORADO OIL-AND-GAS FIELD

As a first example of quadratic regression, I use a small data set from the Rangely, Colorado oil-and-gas field, from an earlier analysis (Koch

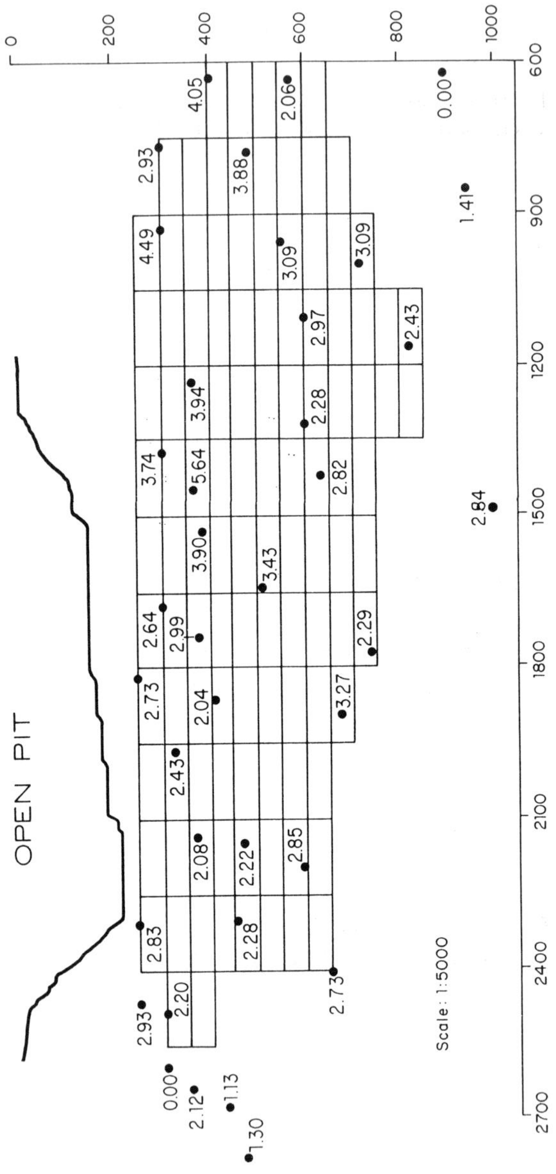

Figure 12.1 Chambishi mine, Zambia, vertical longitudinal section

and Link, 1971, p. 4). The data are elevations in 26 wells of the top of the Weber Sandstone (Pennsylvanian/Permian age) above a datum plane. In worksheet C12A.WK1, Part 1 is a data table listing (in columns C, D, and E) the elevation (w), the east coordinate (x), and the north coordinate (y) for the data plotted in Figure 12.2.

The quadratic equation for a surface in two dimensions is

w-hat = A + (B.1)*X + (B.2)*Y + (B.3)*(X^2) + (B.4)*(Y^2) + (B.5)*(XY),

where A is a constant and B.1 to B.5 are coefficients. Columns F, G, and H in Part 1 list the squared and cross-product terms calculated from columns D and E by entering the appropriate formulas in cells F13, G13, and H13, and then copying downward to the bottom of the table.

Part 2 gives the regression output, which is similar to that of linear regression (Chapter 3) except that there are five rather than two independent variables.

Part 3 applies the coefficients to x,y data points to calculate the estimated values of W-HAT. Plotting the data points, will give you enough to construct the elevation contours of the dome graphed in Figure 12.3.

12.2 QUADRATIC REGRESSION—CHAMBISHI MINE

Worksheet C12B.WK1 is constructed similar to worksheet C12A.WK1. The regression output is similar. The smaller value of r-squared, 0.60, indicates that less of the variability (about 36 percent) is explained by the model; however, I believe that this moderate smoothing may provide appropriate block estimates. Part 3 gives the predicted value of Cu for each block; I arranged the table for ease in copying the center coordinates of each block and the formula.

Figure 12.4 is a 1-2-3 plot of the blocks, scaled approximately to the scale of Figure 12.1 by adjusting the column widths using the /Work-

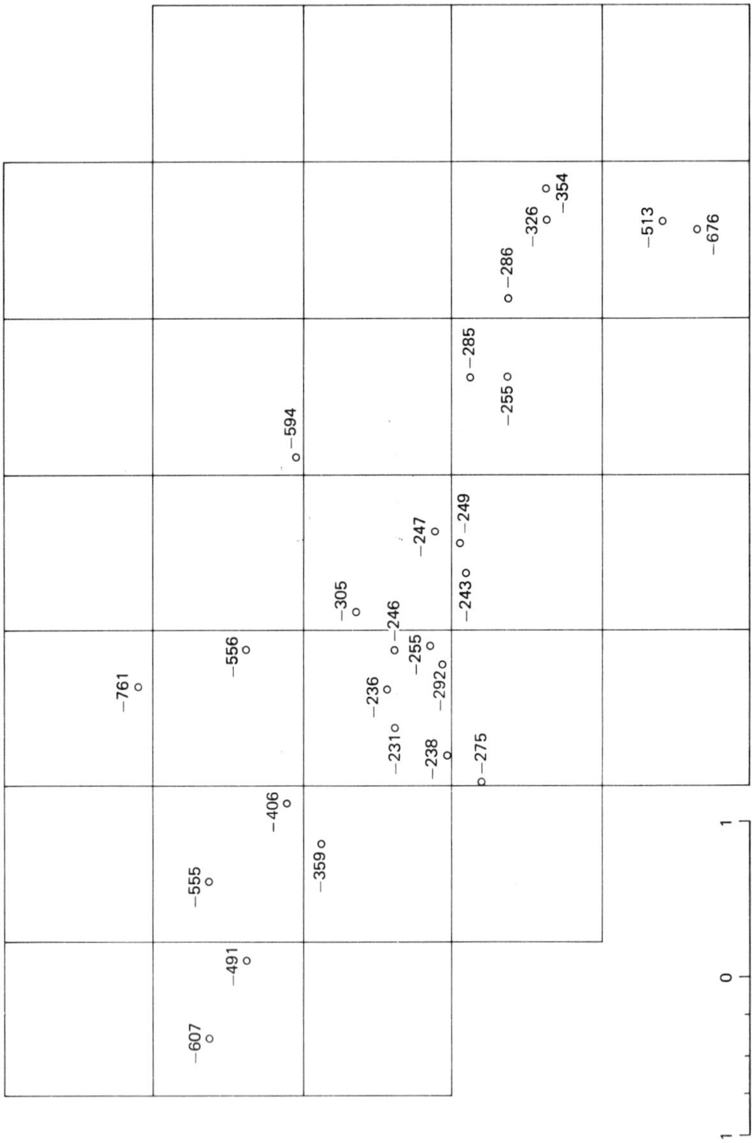

Figure 12.2 Elevations of Weber Sandstone in 29 wells in Rangely oil field, Colorado

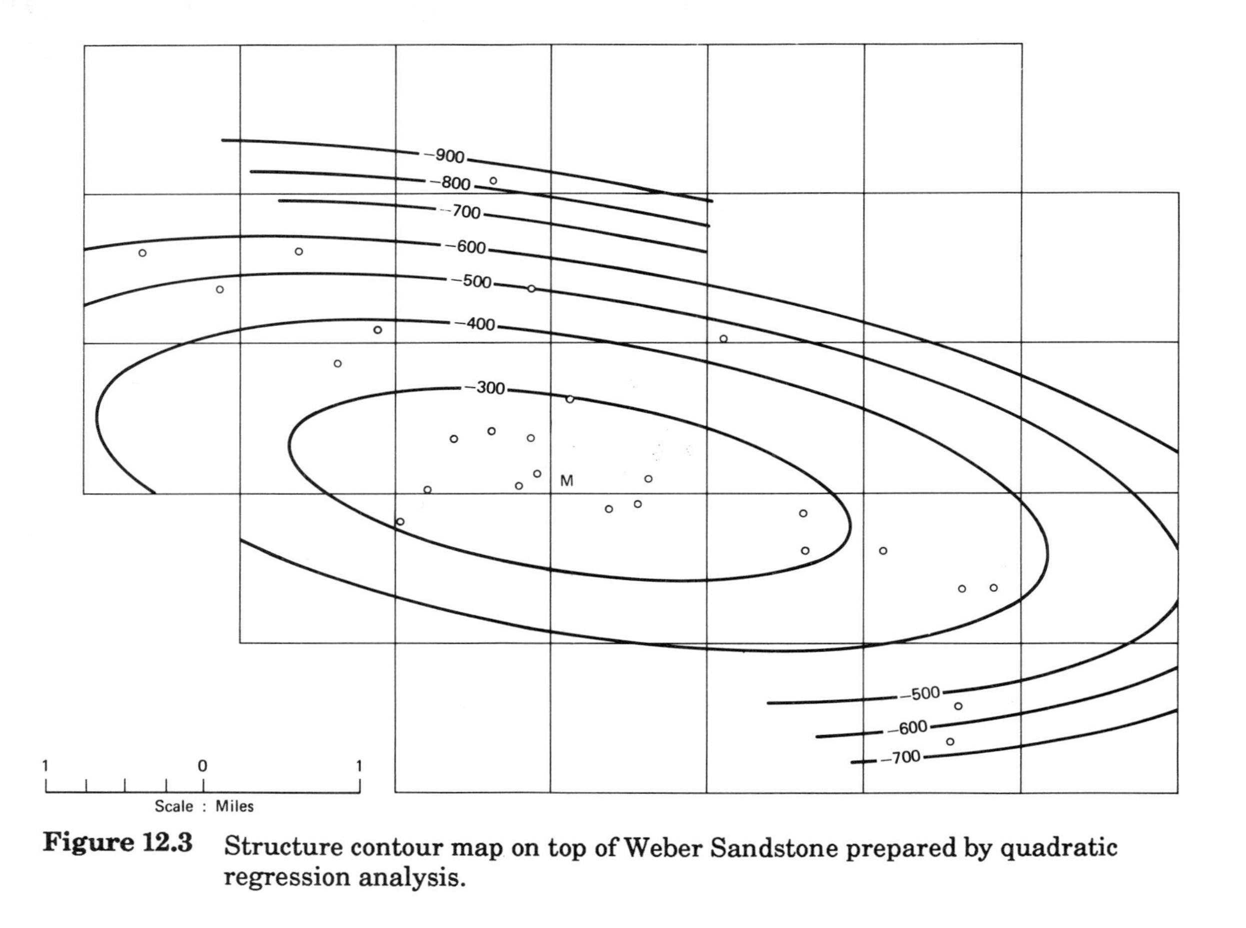

Figure 12.3 Structure contour map on top of Weber Sandstone prepared by quadratic regression analysis.

	2.1	2.5	2.9	3.2	3.5	3.7	3.9	4.1	4.2	4.3		
1.8	2.2	2.5	2.8	3.1	3.4	3.6	3.7	3.9	4.0	4.0	4.0	
1.9	2.2	2.5	2.8	3.1	3.3	3.5	3.6	3.7	3.8	3.8	3.8	
	2.3	2.6	2.8	3.0	3.2	3.3	3.5	3.5	3.5	3.5	3.5	3.4
	2.3	2.6	2.8	3.0	3.1	3.2	3.3	3.4	3.4	3.3	3.2	3.1
	2.4	2.6	2.8	3.0	3.1	3.2	3.2	3.2	3.2	3.1	3.0	2.8
	2.5	2.7	2.8	2.9	3.0	3.1	3.1	3.1	3.0	2.9	2.7	2.6
	2.6	2.7	2.8	2.9	3.0	3.0	3.0	2.9	2.8	2.7	2.5	2.3
				2.9	2.9	2.9	2.9	2.8	2.6	2.5	2.3	
					2.9	2.9	2.8	2.7	2.5	2.3		
								2.5	2.3			
								2.4	2.2			

Figure 12.4 Plot for Chambishi problem

sheet-Global-Width command. To make this plot, I first copied Part 3 to arrange the columns side by side, and then used the command / Transpose to interchange rows and columns.

Part 4 calculates a confidence interval for the population mean, and Part 5 is the table of labels.

12.3 OTHER WAYS TO ESTIMATE BLOCK GRADES

Many other ways have been devised to estimate block grades. Peters (1987, p. 489) explains weighting by the inverse distance squared, and Hayes and Koch (1984) provide a computer program for this method. Table 12.1 gives several grade estimates made in different ways for the Chambishi mine. In a classic paper, Tukey (1948) showed that any reasonable weighting system will yield similar results overall, although values for individual blocks may differ. (Technically, Tukey showed that weights can change much without changing the standard error of an estimate.) Not until the ore is mined out, and then only if reliable mining records have been kept, do we know for any deposit whether a particular evaluation scheme is "best."

You may want to experiment with estimating block values in various ways and entering your results in column E of Part 3 of worksheet C12B.WK1. If you then compute the mean, and confidence intervals, you can compare them with those in Table 12.1. I think that you will be surprised with how close they are, at least this is the experience of students at the University of Georgia through several years.

Table 12.1 Chambishi mine, Zambia. Grade estimates from 38 boreholes made in various ways. Values are in percent copper.

Source of Estimate	Mean	Standard Deviation	90% Confidence Interval	
			Lower	Upper
Arithmetic average of borehole	2.69	1.09	2.39	2.98
Geologist's estimate	2.94	0.38	2.84	3.03
An early computer contouring program	3.03	0.82	2.90	3.15
Quadratic regression	3.00	0.74	2.89	3.12
Inverse distance				
3 neighbors	3.05	0.66	2.94	3.15
4 neighbors	3.03	0.64	2.93	3.13
Inverse distance squared				
3 neighbors	3.05	0.70	2.94	3.16
4 neighbors	3.04	0.69	2.93	3.15

Worksheet C12A Quadratic regression of Rangely data

```
 1   18 Feb 88, file C12A.WK1 on disks GSK 059-061
 2
 3   Worksheet C12A.WK1.  Quadratic regression of the Rangely data
 4
 5   Part 1. Table of data for the 26 wells.  W, X, and Y are from
 6   the original data file (Koch and Link, 1971, p. 34); X**2,
 7   Y**2, and XY are computed as explained in the text.
 8
 9      C         D         E         F         G         H
10    ----------------------------------------------------------
11      W         X         Y       X**2      Y**2       XY
12    ----------------------------------------------------------
13       607      4.35      3.64   18.9225   13.2496   15.8340
14       491      4.87      3.38   23.7169   11.4244   16.4606
15       555      5.38      3.64   28.9444   13.2496   19.5832
16       406      5.88      3.10   34.5744    9.6100   18.2280
17       556      6.86      3.36   47.0596   11.2896   23.0496
18       761      6.62      4.10   43.8244   16.8100   27.1420
19       594      8.10      3.02   65.6100    9.1204   24.4620
20       359      5.61      2.86   31.4721    8.1796   16.0446
21       231      6.36      2.38   40.4496    5.6644   15.1368
22       236      6.61      2.43   43.6921    5.9049   16.0623
23       246      6.87      2.38   47.1969    5.6644   16.3506
24       238      6.20      2.01   38.4400    4.0401   12.4620
25       292      6.80      2.05   46.2400    4.2025   13.9400
26       255      6.90      2.13   47.6100    4.5369   14.6970
27       305      7.16      2.62   51.2656    6.8644   18.7592
28       247      7.64      2.10   58.3696    4.4100   16.0440
29       275      6.02      1.81   36.2404    3.2761   10.8962
```

Worksheet C12A *(continued)*

```
30        243      7.38      1.90  54.4644    3.6100  14.0220
31        249      7.56      1.94  57.1536    3.7636  14.6664
32        255      8.63      1.63  74.4769    2.6569  14.0669
33        285      8.63      1.88  74.4769    3.5344  16.2244
34        286      9.13      1.62  83.3569    2.6244  14.7906
35        326      9.63      1.37  92.7369    1.8769  13.1931
36        354      9.84      1.37  96.8256    1.8769  13.4808
37        676      9.59      0.35  91.9681    0.1225   3.3565
38        513      9.62      0.60  92.5444    0.3600   5.7720
39   -----------------------------------------------------
40
41  Part 2.   Regression output
42
43   -----------------------------------------------------
44  Constant                         2628.752
45  Std Err of Y Est                 47.59915
46  R Squared                        0.928588
47  No. of Observations                    26
48  Degrees of Freedom                     20
49
50   X Coefficient(s)   -392.922 -968.661 20.72697 153.8660 48.16129
51   Std Err of Coef.    204.040 270.2769 10.29091 19.93190 25.20694
52
53
```

Worksheet C12A *(continued)*

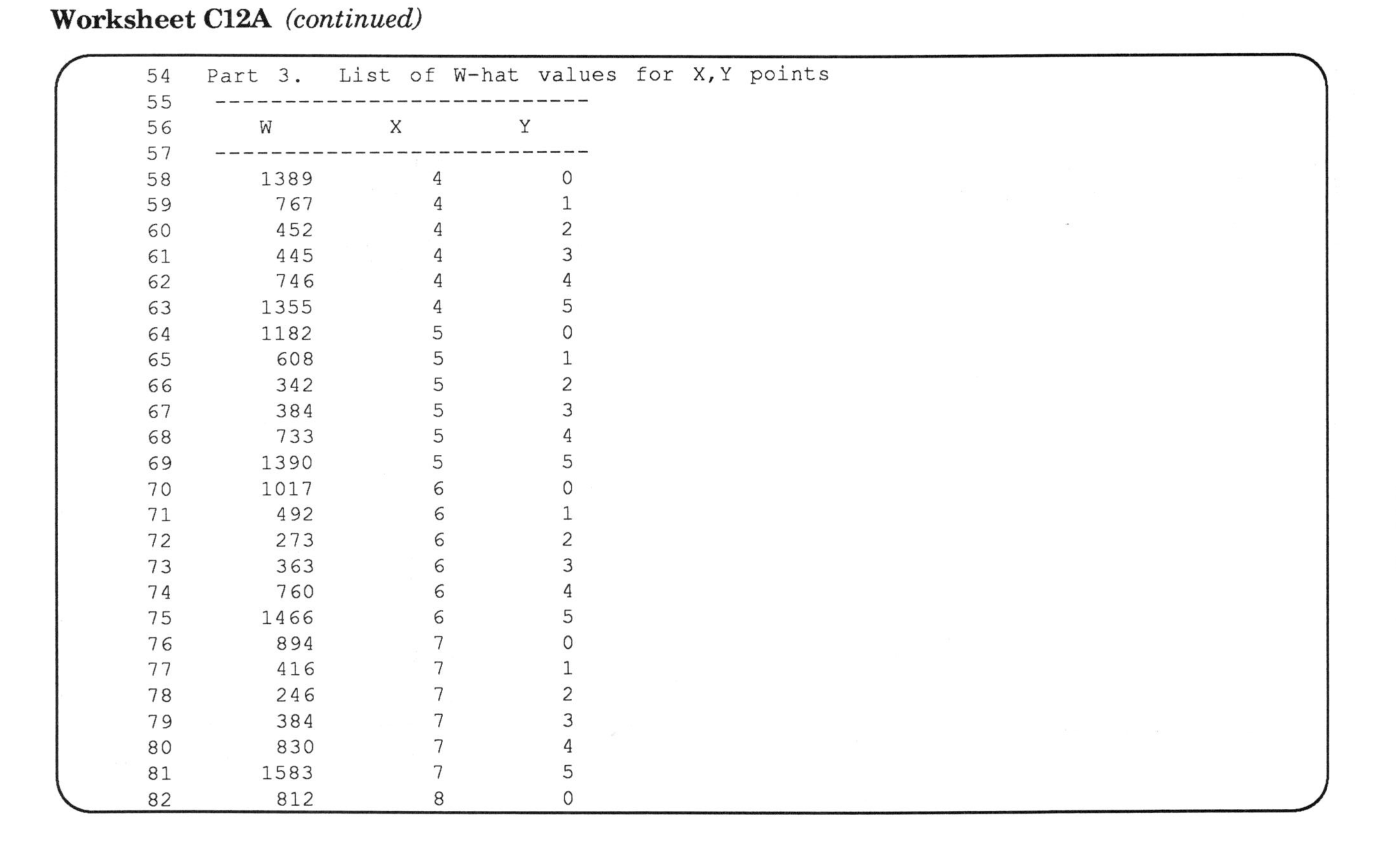

54	Part 3. List of W-hat values for X,Y points		
55	----------------------------		
56	W	X	Y
57	----------------------------		
58	1389	4	0
59	767	4	1
60	452	4	2
61	445	4	3
62	746	4	4
63	1355	4	5
64	1182	5	0
65	608	5	1
66	342	5	2
67	384	5	3
68	733	5	4
69	1390	5	5
70	1017	6	0
71	492	6	1
72	273	6	2
73	363	6	3
74	760	6	4
75	1466	6	5
76	894	7	0
77	416	7	1
78	246	7	2
79	384	7	3
80	830	7	4
81	1583	7	5
82	812	8	0

Worksheet C12A *(continued)*

```
 83       382       8       1
 84       261       8       2
 85       447       8       3
 86       940       8       4
 87      1742       8       5
 88       771       9       0
 89       390       9       1
 90       316       9       2
 91       551       9       3
 92      1092       9       4
 93      1942       9       5
 94       772      10       0
 95       439      10       1
 96       414      10       2
 97       696      10       3
 98      1286      10       4
 99      2184      10       5
100       815      11       0
101       530      11       1
102       552      11       2
103       883      11       3
104      1521      11       4
105      2467      11       5
106    -----------------------------
107
108
```

Worksheet C12A *(continued)*

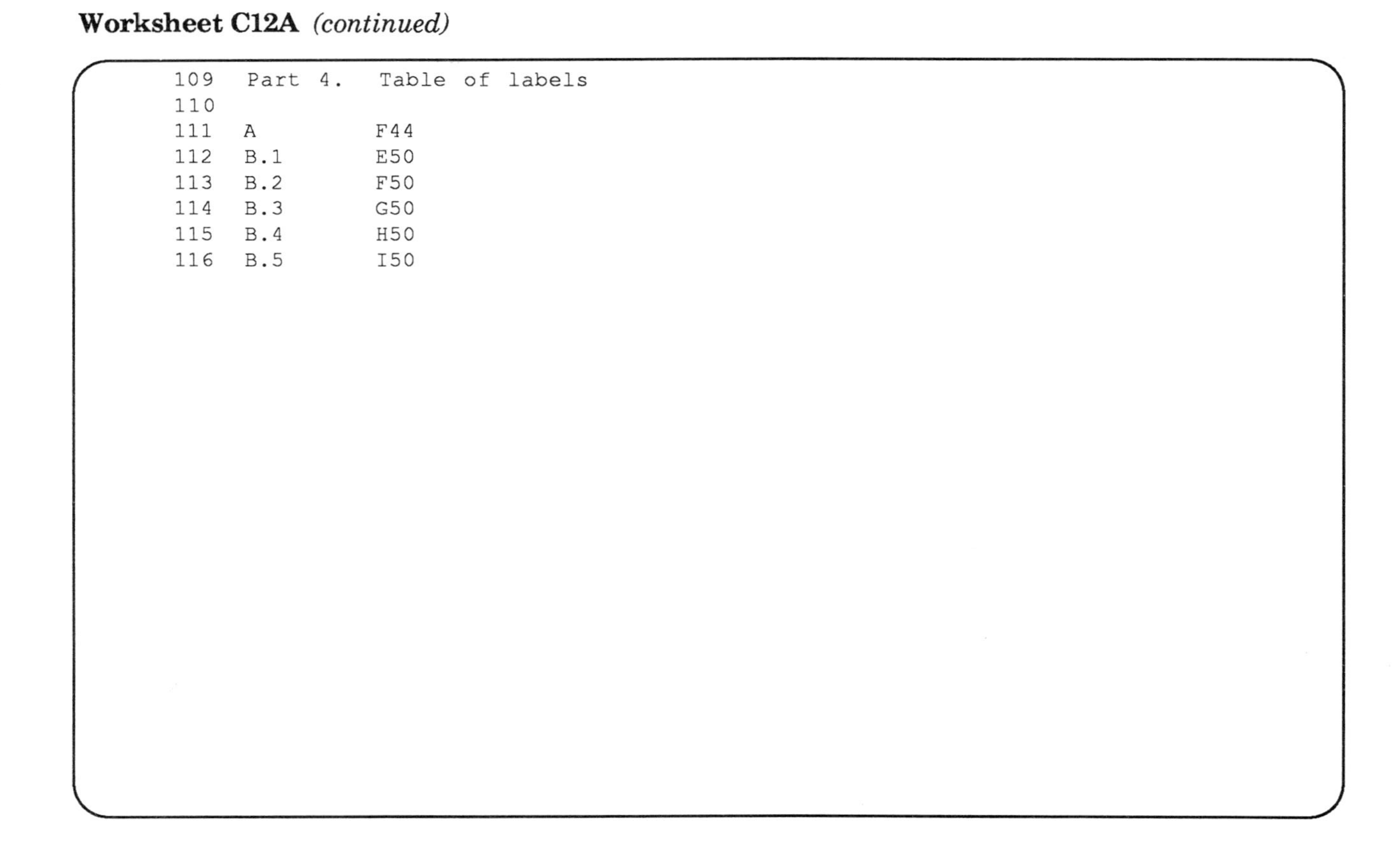

```
109   Part 4.   Table of labels
110
111   A         F44
112   B.1       E50
113   B.2       F50
114   B.3       G50
115   B.4       H50
116   B.5       I50
```

Worksheet C12B Quadratic regression of Chambishi data

```
1    22 Feb 88, file C12B.WK1 on disks GSK 059-061
2
3    Worksheet C12B.WK1.  Quadratic regression of the Chambishi data
4
5    Part 1.  Table of data for the 38 drillholes.  Cu, X, and Y
6    values from the original records; X**2, Y**2, and XY are
7    computed as explained in the text.
8
9        C          D          E         F          G          H
10    ----------------------------------------------------------
11      Cu          X          Y        X**2       Y**2        XY
12    ----------------------------------------------------------
13      4.05       632        412    399,424    169,744    260,384
14      2.06       636        577    404,496    332,929    366,972
15      2.93       763        296    582,169     87,616    225,848
16      3.88       777        485    603,729    235,225    376,845
17      4.49       928        302    861,184     91,204    280,256
18      3.09       955        555    912,025    308,025    530,025
19      3.09      1000        720  1,000,000    518,400    720,000
20      0.00       625        900    390,625    810,000    562,500
21      1.41       855        945    731,025    893,025    807,975
22      2.97      1105        600  1,221,025    360,000    663,000
23      2.43      1165        820  1,357,225    672,400    955,300
24      3.94      1230        368  1,512,900    135,424    452,640
25      2.28      1310        600  1,716,100    360,000    786,000
26      3.74      1375        298  1,890,625     88,804    409,750
27      2.82      1427        635  2,036,329    403,225    906,145
28      5.64      1450        365  2,102,500    133,225    529,250
29      3.90      1530        385  2,340,900    148,225    589,050
```

Worksheet C12B *(continued)*

30	2.84	1490	995	2,220,100	990,025	1,482,550
31	3.43	1645	510	2,706,025	260,100	838,950
32	2.64	1675	297	2,805,625	88,209	497,475
33	2.99	1735	375	3,010,225	140,625	650,625
34	2.29	1772	740	3,139,984	547,600	1,311,280
35	2.73	1822	250	3,319,684	62,500	455,500
36	2.04	1865	410	3,478,225	168,100	764,650
37	3.27	1895	675	3,591,025	455,625	1,279,125
38	2.43	1975	327	3,900,625	106,929	645,825
39	2.08	2140	370	4,579,600	136,900	791,800
40	2.22	2147	465	4,609,609	216,225	998,355
41	2.85	2195	595	4,818,025	354,025	1,306,025
42	2.28	2303	450	5,303,809	202,500	1,036,350
43	2.731	2400	650	5,760,000	422,500	1,560,000
44	2.83	2305	250	5,313,025	62,500	576,250
45	2.93	2475	240	6,125,625	57,600	594,000
46	2.20	2475	305	6,125,625	93,025	754,875
47	0.00	2592	300	6,718,464	90,000	777,600
48	2.12	2635	357	6,943,225	127,449	940,695
49	1.13	2670	430	7,128,900	184,900	1,148,100
50	1.30	2775	470	7,700,625	220,900	1,304,250
51	----------	----------	----------	----------	----------	----------
52						
53						

Worksheet C12B *(continued)*

```
54   Part 2.   Regression Output
55   ------------------------------------------------------------
56   Constant                              6.6027
57   Std Err of Y Est                      0.7406
58   R Squared                               0.60
59   No. of Observations                       38
60   Degrees of Freedom                        32
61
62   X Coefficient(s)   -3.48E-05 -1.01E-02 -8.27E-07  1.52E-06 4.28E-06
63   Std Err of Coef.    1.50E-03  5.04E-03  3.43E-07  3.25E-06 1.36E-06
64   ------------------------------------------------------------
65
66
67   Part 3.   List of Cu values for each block
68   ----------------------------------------
69     Cu         x         y
70   ----------------------------------------
71      2.13      2325       275
72      2.51      2175       275
73      2.86      2025       275
74      3.18      1875       275
75      3.45      1725       275
76      3.69      1575       275
77      3.89      1425       275
78      4.05      1275       275
79      4.18      1125       275
80      4.27       975       275
81      1.77      2475       325
82      2.16      2325       325
```

Worksheet C12B *(continued)*

83	2.52	2175	325
84	2.84	2025	325
85	3.12	1875	325
86	3.36	1725	325
87	3.57	1575	325
88	3.74	1425	325
89	3.87	1275	325
90	3.96	1125	325
91	4.02	975	325
92	4.04	825	325
93	1.85	2475	375
94	2.21	2325	375
95	2.53	2175	375
96	2.82	2025	375
97	3.07	1875	375
98	3.28	1725	375
99	3.45	1575	375
100	3.59	1425	375
101	3.69	1275	375
102	3.75	1125	375
103	3.78	975	375
104	3.76	825	375
105	2.26	2325	425
106	2.56	2175	425
107	2.81	2025	425
108	3.03	1875	425
109	3.20	1725	425
110	3.35	1575	425
111	3.45	1425	425

Worksheet C12B *(continued)*

112	3.52	1275	425
113	3.55	1125	425
114	3.54	975	425
115	3.50	825	425
116	3.42	675	425
117	2.33	2325	475
118	2.58	2175	475
119	2.81	2025	475
120	2.99	1875	475
121	3.14	1725	475
122	3.25	1575	475
123	3.32	1425	475
124	3.36	1275	475
125	3.35	1125	475
126	3.31	975	475
127	3.24	825	475
128	3.12	675	475
129	2.39	2325	525
130	2.62	2175	525
131	2.81	2025	525
132	2.96	1875	525
133	3.08	1725	525
134	3.16	1575	525
135	3.20	1425	525
136	3.20	1275	525
137	3.17	1125	525
138	3.09	975	525
139	2.99	825	525
140	2.84	675	525

Worksheet C12B *(continued)*

141	2.47	2325	575
142	2.67	2175	575
143	2.82	2025	575
144	2.94	1875	575
145	3.03	1725	575
146	3.07	1575	575
147	3.08	1425	575
148	3.05	1275	575
149	2.99	1125	575
150	2.88	975	575
151	2.74	825	575
152	2.56	675	575
153	2.56	2325	625
154	2.72	2175	625
155	2.84	2025	625
156	2.93	1875	625
157	2.98	1725	625
158	3.00	1575	625
159	2.97	1425	625
160	2.91	1275	625
161	2.81	1125	625
162	2.68	975	625
163	2.51	825	625
164	2.30	675	625
165	2.93	1875	675
166	2.95	1725	675
167	2.93	1575	675
168	2.87	1425	675
169	2.78	1275	675

Worksheet C12B *(continued)*

```
170      2.65     1125       675
171      2.48      975       675
172      2.28      825       675
173      2.92     1725       725
174      2.87     1575       725
175      2.78     1425       725
176      2.65     1275       725
177      2.49     1125       725
178      2.29      975       725
179      2.54     1275       775
180      2.34     1125       775
181      2.43     1275       825
182      2.20     1125       825
183    ----------------------------------------
184
185
186   Part 4.  Calculation of a confidence interval for the population
187       mean (Notation from Worksheet C2A.WK1)
188    ----------------------------------------
189   Value                Item
190    ----------------------------------------
191        112             n
192
193       3.00             Mean
194
195       0.74              Standard deviation
196
197      1.658             t(5%)
198
```

Worksheet C12B *(continued)*

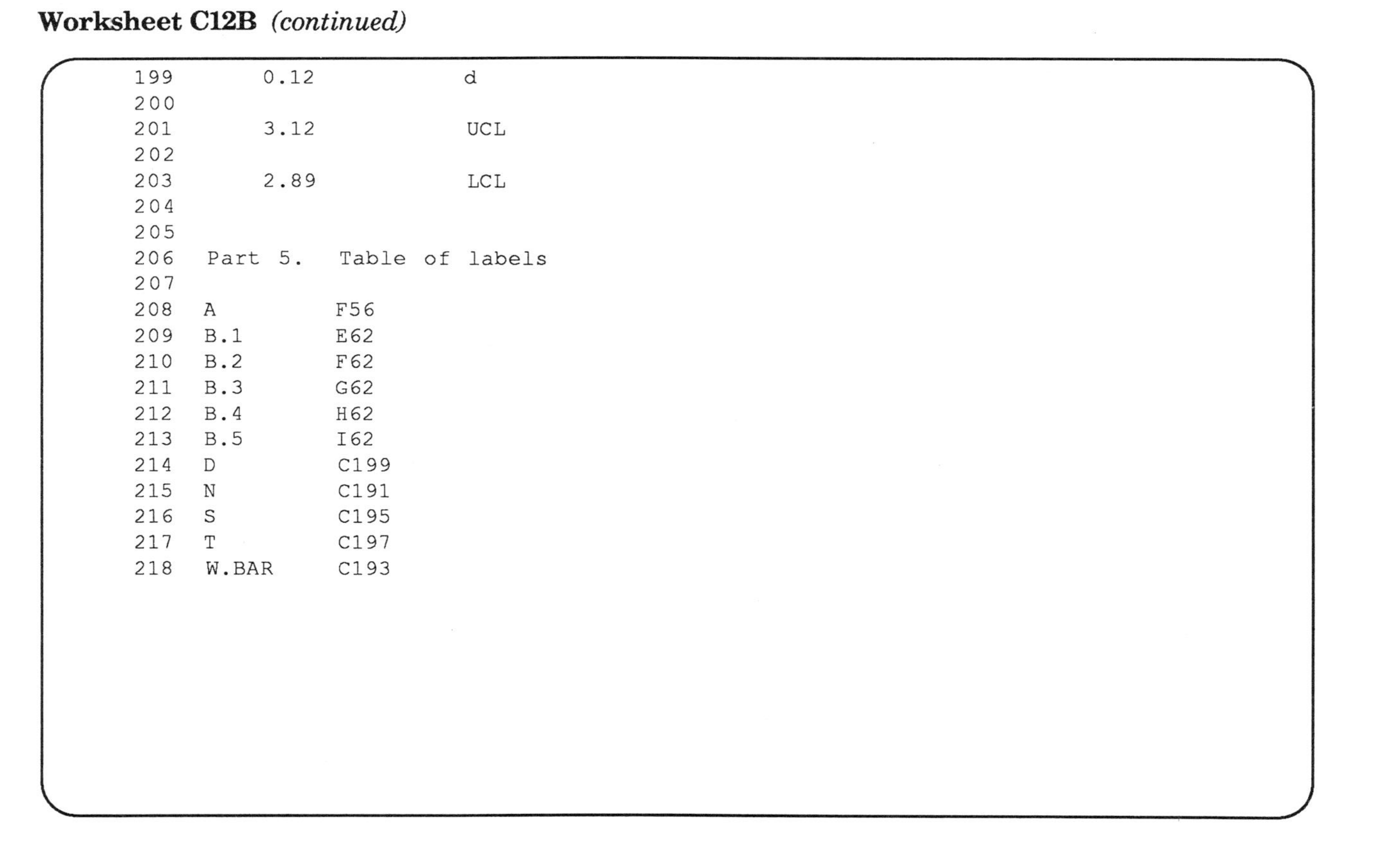

```
199      0.12          d
200
201      3.12          UCL
202
203      2.89          LCL
204
205
206  Part 5.  Table of labels
207
208  A        F56
209  B.1      E62
210  B.2      F62
211  B.3      G62
212  B.4      H62
213  B.5      I62
214  D        C199
215  N        C191
216  S        C195
217  T        C197
218  W.BAR    C193
```

CHAPTER 13

Ore Concentration and Smelter Settlement

The worksheets in this chapter calculate the concentration of ore after mining and the payments made by smelters for the concentrated products. Two of the worksheets are taken from the comprehensive model in the next chapter; they are separated here so that you can easily work with and modify them.

13.1 ORE CONCENTRATION

After mining, ores are generally separated in a plant termed a mill or a concentrator into two types of products: one or more concentrates containing most of the valuable minerals, and tailings containing most of the other minerals. Because the minerals are processed in an industrial operation, their recovery is imperfect. A materials balance accounts for the ore, named the heads, entering the mill, and the concentrates and tailings leaving the mill.

Worksheet C13A.WK1 is a simplified materials balance for the concentrator at the El Indio mine in Chile (Smith, 1986). It is convenient to consider the concentration of 100 tonnes of 5 percent copper ore; these numbers are entered in cells D12 and E12. Multiplying tonnes of heads (D12) by grade (E12) gives the contents, in tonnes of copper (F12); we can express this fundamental equation as

$$T*G = C,$$

where T is tonnes, G is grade, and C is the quantity (also named the contents) of contained metal. Multiplying F12 by the percentage recovery of copper (94 percent) gives the contents of copper in the concentrate in F13. Because the concentrate grade is more or less a constant, being largely a function of the efficiency of the concentrator, the tonnes of concentrate in D13 is calculated by dividing F13 by E13 (with a constant to express percentage). Finally, the tonnes and copper contents of the tailings are determined by subtracting the concentrate values from the heads values, and the tailings grade is calculated from the fundamental equation.

Worksheet C13B.WK1 is similar to the previous worksheet with the inclusion of gold, which was omitted for simplicity from the first worksheet. The gold adds no significant additional tonnage to the concentrates.

Worksheet C13C.WK1 is the materials balance that Peters (1987, p. 252) uses for his comprehensive model; the original data, for the Bunker Hill mine in Idaho, are from Sather and Prindle (1970). Two concentrates are produced, a lead concentrate containing most of the lead and silver, and a zinc concentrate containing most of the zinc and some of the silver. In this worksheet, the rows and columns are transposed for convenience in printing. I assumed that the percentages of lead and zinc in the concentrates and the percentage recoveries of all three metals are constants. Therefore, the concentrate weights are linear functions of the grades of lead and zinc in the mined ore. Other assumptions are possible, but making them would require a detailed knowledge of the particular mineralogy and the concentration scheme.

13.2 CALCULATION OF GRADE FOR A MINED-OUT DEPOSIT

An interesting use of the materials balance in the previous section combines it with Section 11.2 on calculating the specific gravity of ore. Here is a problem for you to consider.

In examining the Veta Grande lead mine in Mexico, you learn from old maps, checked by your observations and measurements, that about 10,000 cubic meters of ore was mined. The record of concentrates shipped is as follows:

Shipments	Tonnes	Assay
Lead concentrate	4,000	65% Pb

The problem is to estimate the grade of ore as mined, because there is a potential for additional ore of a similar grade at depth.

To simplify the problem, we may assume that (1) the lead concentrate consisted entirely of galena plus pyrite, (2) the gangue rejected in milling consisted entirely of limestone, and (3)the recovery of metals was perfect.

You have at hand data on the specific gravities of the minerals and limestone, and the atomic weights of the minerals.

Worksheet C13D.WK1 solves this problem. Part 1 provides constants; the percent of lead in galena (row 13) is obtained from the atomic weights for lead and sulfur.

Part 2 does the calculations. Working from the given data, we first determine the tonnes of ore mined (row 27), equivalent to a specific gravity of 2.89, and then the lead grade (row 29).

This worksheet is an example of problems usually encountered in economic geology. Generally, the problem resolves itself into selecting the pertinent data and using the basic materials balance relations to determine the required solution. For instance, we might need to work backward from the tailings to determine tonnage and grade of ore as mined.

13.3 SMELTER SETTLEMENT

Concentrates are sold to a smelter, according to negotiated contracts. Peters (1987, p. 255-256) provides "abbreviated examples of typical 'open' smelter schedules, the type furnished to independent shippers of ores and concentrates by custom metallurgical plants." Peters does the calculations (1987, p. 257) for the Bunker Hill ore of worksheet C13C.WK1. His calculations, which are self-explanatory, are given in worksheet C13E.WK1.

Worksheet C13A Materials balance for El Indio mine

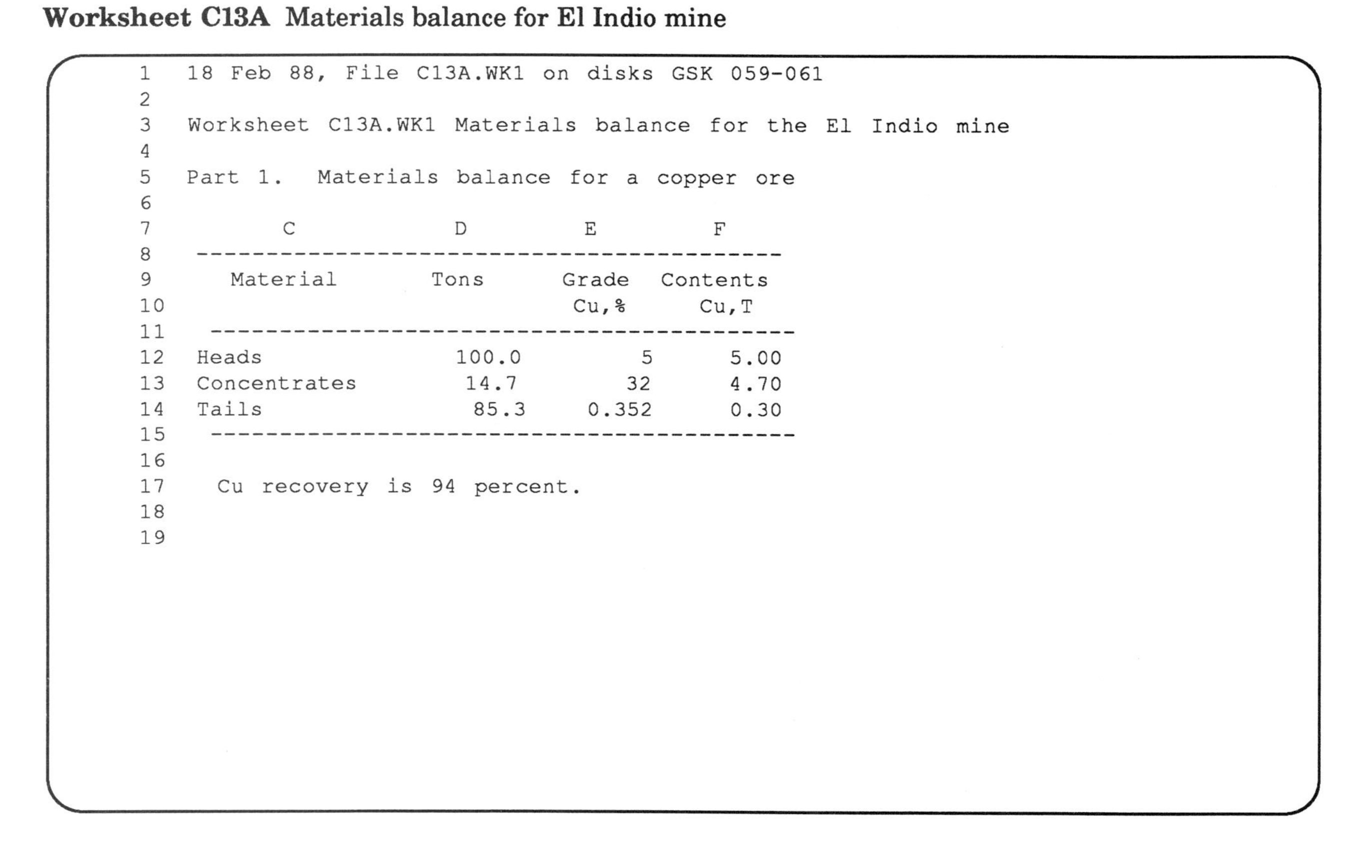

```
1   18 Feb 88, File C13A.WK1 on disks GSK 059-061
2
3   Worksheet C13A.WK1 Materials balance for the El Indio mine
4
5   Part 1.  Materials balance for a copper ore
6
7          C            D         E         F
8    -----------------------------------------
9      Material       Tons      Grade  Contents
10                               Cu,%      Cu,T
11    -----------------------------------------
12   Heads             100.0         5      5.00
13   Concentrates       14.7        32      4.70
14   Tails              85.3     0.352      0.30
15    -----------------------------------------
16
17    Cu recovery is 94 percent.
18
19
```

Worksheet C13A *(continued)*

```
20  Part 2.  Table of labels
21
22  CON.TONS          D13
23  CU.%.IN.CON       E13
24  CU.%.IN.HEADS     E12
25  CU.%.IN.TAILS     F14
26  CU.T.IN.CON       F13
27  CU.T.IN.HEADS     F12
28  CU.T.IN.TAILS     F14
29  HEADS.TONS        D12
30  TAILS.TONS        D14
```

Worksheet C13B Materials balance for El Indio mine with gold included

```
 1   18 Feb 88, File C13B.WK1 on disks GSK 059-061 2
 3   Worksheet C13B.WK1.  Materials balance for the El Indio mine
 4        with gold included
 5
 6
 7   Part 1.  Materials balance for a copper, gold ore
 8
 9        C             D          E          F          G          H
10    -----------------------------------------------------------------
11      Material       Tons      Grade      Grade   Contents  Contents
12                              Au,g/T      Cu,%      Au,g      Cu,T
13    -----------------------------------------------------------------
14   Heads             100.0        18       5.00      1800         5
15   Concentrates       14.7        83      32.00      1278       4.7
16   Tails              85.3     6.119       0.35       522       0.3
17    -----------------------------------------------------------------
18
19   Au recovery is 71 %.  Cu recovery is 94 %.
20
21
```

Worksheet C13A *(continued)*

```
22   Part 2.   Table of labels
23
24   AU.G.IN.CON      G15
25   AU.G.IN.HEADS    G14
26   AU.G.IN.TAILS    G16
27   AU.G.P.T.CON     E15
28   AU.G.P.T.HEADS   E14
29   AU.G.P.T.TAILS   E16
30   CON.TONS         D15
31   CU.%.IN.CON      F15
32   CU.%.IN.HEADS    F14
33   CU.%.IN.TAILS    F16
34   CU.T.IN.CON      H15
35   CU.T.IN.HEADS    H14
36   CU.T.IN.TAILS    H16
37   HEADS.TONS       D14
38   TAILS.TONS       D16
```

Worksheet C13C Materials balance for Bunker Hill mine

```
 1  22 Feb 88, file C13C.WK1 on disks GSK 059-061
 2
 3  Worksheet C13C.WK1.  Materials balance for the Bunker Hill mine
 4
 5  Part 1.  Detail of materials balance
 6    (from Peters, 1987, p. 252)
 7                                     D         E         F         G
 8   ----------------------------------------------------------------------
 9                                    Pb        Zn       Tail-     Heads
10                                    Conc.     Conc.    ings
11   ----------------------------------------------------------------------
12   Assays
13     Percent weight                   8.84      7.85     83.31    100.00
14     Ag, oz/mt                       44.57      3.13      0.19      4.34
15     Ag, g/mt                      1385.99     97.25      6.05     135.1
16     Pb, %                           66.59      1.18      0.25      6.18
17     Zn, %                            4.98     54.21      0.25      4.90
18
19   Contents
20     Ag, grams                                           504.0   13510.0
21     Ag, grams in Pb concentrate  12245.5
22     Ag, grams in Zn concentrate              763.5
23     Pb, kilograms                   588.3       9.3      20.7    6659.0
24     Zn, kilograms                    44.0     425.6      20.8     498.0
25
26   Percentage distribution
27     Ag                              90.62      5.65      3.73    100.00
28     Pb                              95.16      1.50      3.34    100.00
29     Zn                               8.97     86.78      4.25    100.00
```

Worksheet C13C *(continued)*

```
30   ------------------------------------------------------------------
31
32
33
34   Part 2.  Table of labels
35
36   %.WT.PB.CON                     D13
37   %.WT.TAILS                      F13
38   %.WT.ZN.CON                     E13
39   AG.G.IN.HEADS                   G20
40   AG.G.IN.PB.CON                  D21
41   AG.G.IN.TAILS                   F20
42   AG.G.IN.ZN.CON                  E22
43   AG.G.P.T.HEADS                  G15
44   AG.G.P.T.PB.CON                 D15
45   AG.G.P.T.TAILS                  F15
46   AG.G.P.T.ZN.CON                 E15
47   AG.REC.PB.CON                   D27
48   AG.REC.ZN.CON                   E27
49   HEADS.TONS                      G13
50   PB.%.IN.HEADS                   G16
51   PB.%.IN.PB.CON                  D16
52   PB.%.IN.TAILS                   F23
53   PB.%.IN.ZN.CON                  E16
54   PB.%.REC.PB.CON                 D28
55   PB.%.REC.ZN.CON                 E28
56   PB.KG.IN.HEADS                  G23
57   PB.KG.IN.PB.CON                 D23
58   PB.KG.IN.ZN.CON                 E23
```

Worksheet C13C *(continued)*

```
59   ZN.%.IN.HEADS                     G17
60   ZN.%.IN.PB.CON                    D17
61   ZN.%.IN.ZN.CON                    E17
62   ZN.%.REC.PB.CON                   D29
63   ZN.%.REC.ZN.CON                   E29
64   ZN.KG.IN.HEADS                    G24
65   ZN.KG.IN.PB.CON                   D24
66   ZN.KG.IN.TAILS                    F24
67   ZN.KG.IN.ZN.CON                   E24
```

Worksheet C13D Estimating lead grade for prospect

```
 1  18 Feb 88, file C13D.WK1 on disks GSK 059-061
 2
 3  Worksheet C13D.WK1.  Estimating lead grade for a prospect
 4
 5  Part 1. Constants
 6  ---------------------------------------------------------
 7  Lead, % in concentrate                    65
 8  Lead, atomic weight                    207.2
 9  Sulfur, atomic weight                  32.06
10  Lead, % in galena                       86.6
11  Galena, specific gravity                 7.5
12  Pyrite, specific gravity                 5.0
13  Limestone, specific gravity             2.65
14  ---------------------------------------------------------
15
16
17  Part 2.  Calculations
18  ---------------------------------------------------------
19            Material                   Tons    Cubic Meters
20  ---------------------------------------------------------
21  Volume of ore mined                              10,000
22  Lead concentrate produced           4,000
23  Lead                                2,600
24  Galena                              3,002           400
25  Pyrite                                998           200
26  Limestone                          24,910         9,400
27  Weight of ore mined                28,910
28  ---------------------------------------------------------
29  Grade of lead                        8.99
```

Worksheet C13D *(continued)*

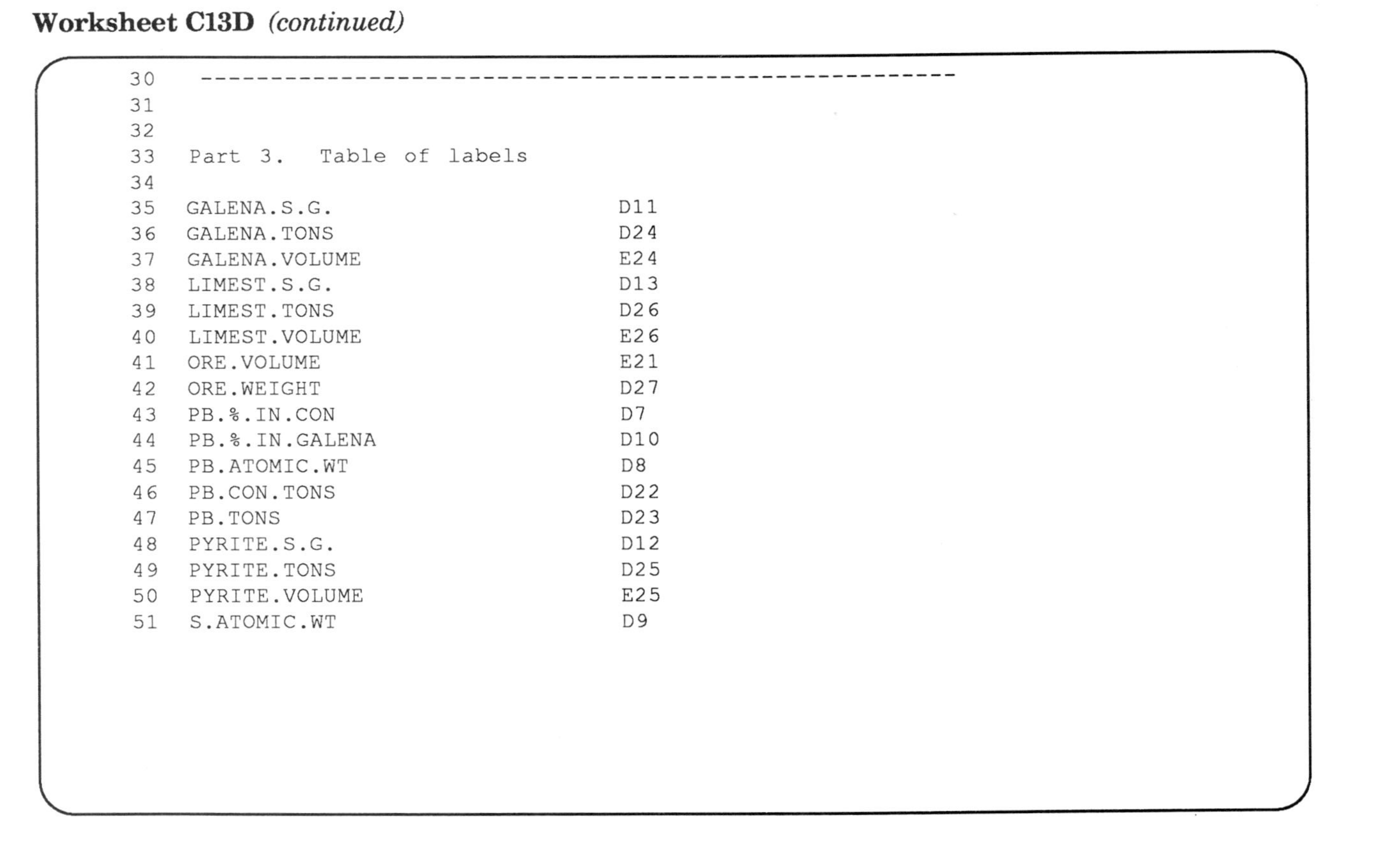

```
30   ----------------------------------------------------------
31
32
33   Part 3.   Table of labels
34
35   GALENA.S.G.                        D11
36   GALENA.TONS                        D24
37   GALENA.VOLUME                      E24
38   LIMEST.S.G.                        D13
39   LIMEST.TONS                        D26
40   LIMEST.VOLUME                      E26
41   ORE.VOLUME                         E21
42   ORE.WEIGHT                         D27
43   PB.%.IN.CON                        D7
44   PB.%.IN.GALENA                     D10
45   PB.ATOMIC.WT                       D8
46   PB.CON.TONS                        D22
47   PB.TONS                            D23
48   PYRITE.S.G.                        D12
49   PYRITE.TONS                        D25
50   PYRITE.VOLUME                      E25
51   S.ATOMIC.WT                        D9
```

Worksheet C13E Smelter settlement

```
 1   18 Feb 88, file C13.E.WK1 on disks GSK 059-061
 2
 3   Worksheet C13E.WK1.   Smelter settlement
 4
 5   Part 1.   Concentrate grades (from worksheet C13C.WK1)
 6
 7                    C                     D        E        F
 8   -------------------------------------------------------------
 9               Material                Ag,g/T    Pb,%     Zn,%
10   -------------------------------------------------------------
11   Lead concentrate                    1386.0     66.6      5.0
12   Zinc concentrate                      97.3      1.2     54.2
13   -------------------------------------------------------------
14
15
```

Worksheet C13E *(continued)*

```
16   Part 2.   Smelter settlements
17      (from Peters, 1987, p. 257)
18
19                       C                       D
20    ---------------------------------------------
21                      Item                Example
22    ---------------------------------------------
23   Metal prices
24     Silver
25       $ per troy ounce                      6.50
26       $ per gram                            0.21
27     Lead
28       $ per pound                           0.20
29       $ per kilogram                        0.44
30
31     Zinc
32       $ per pound                           0.45
33       $ per kilogram                        0.99
34
35   I.   Lead concentrate
36    ---------------------------------------------
37   Moisture, %                                  5
38   Freight charge, $/tonne                  10.00
39   Freight charge, $/dry tonne              10.53
40
41   Payments                                472.78
42     Silver                                261.93
43     Lead                                  210.86
44
```

Worksheet C13E *(continued)*

```
45  Deductions
46    Treatment charge                         70.00
47    Zn penalty                                0.00
48
49  Value at mine, per dry tonne              392.26
50
51
52  II.  Zinc concentrate
53   --------------------------------------------------
54  Moisture, %                                 6.00
55  Freight charge, $/tonne                    12.00
56  Freight charge, $/dry tonne                12.77
57
58  Payments                                  420.64
59    Silver                                   10.27
60    Lead                                      0.00
61    Zinc                                    410.37
62
63  Deductions
64    Treatment charge                        156.00
65
66  Value at mine, per dry tonne              251.88
67
68
```

Worksheet C13E *(continued)*

```
69   Part 3.  Table of labels
70
71   AG.!.PER.GRAM                        D26
72   AG.G.P.T.PB.CON                      D11
73   AG.G.P.T.ZN.CON                      D12
74   AG.PAY.PB.CON                        D42
75   AG.PAY.ZN.CON                        D59
76   FREIGHT.PB.CON                       D39
77   FREIGHT.ZN.CON                       D56
78   PB.!.PER.KILO                        D29
79   PB.%.IN.PB.CON                       E11
80   PB.%.IN.ZN.CON                       E12
81   PB.CON.PRICE                         D49
82   PB.CON.TREAT                         D46
83   PB.PAY.PB.CON                        D43
84   PB.PAY.ZN.CON                        D60
85   TOT.PAY.PB.CON                       D41
86   TOT.PAY.ZN.CON                       D58
87   ZN.!.PER.KILO                        D33
88   ZN.%.IN.PB.CON                       F11
89   ZN.%.IN.ZN.CON                       F12
90   ZN.CON.TREAT                         D64
91   ZN.PAY.ZN.CON                        D61
92   ZN.PENALTY                           D47
```

CHAPTER 14

Peters's Model for Mineral Property Evaluation

As a financial model for mine evaluation, I use the one presented by Peters (1987, p. 568-570) in a table entitled "Steps in estimating profitability of an undeveloped mineral property," together with accompanying tables from Peters's book. This model presents up-to-date thinking of an expert as well as recent costs.

The Lotus 1-2-3 worksheet for the model is file C14A.WK1. Except for round-off errors, the results match those of Peters.

14.1 THE BASIC MODEL

The basic 1-2-3 model is in four parts, corresponding with Peters's format. These four parts are self-explanatory.

Part 5 is a table of cash flows derived from the cash flows for years 1 to 3, labeled !CASH.FLOW.1, !CASH.FLOW.2, and !CASH.FLOW.3, from row 99. The cash flows for years 4 to 14 are the same as that for year 3, because Peters makes no provision for return of working capital in year 14. Years 15 to 20 are provided for your convenience in changing the model, if you wish. These cash flows define the label, ANN.CASH.FLOWS, which is the range D136..D155, used to calculate DCFROR, in cell D99.

Part 6 is the materials balance table from Peters (1987, p. 252). In constructing this table, I assumed that the percentages of lead and zinc in the concentrates and the percentage recoveries of all three metals are constants. Therefore, the concentrate weights are taken as linear functions of the grade of lead and zinc in the ore as mined. Other assumptions are possible.

Part 6 is the calculation of percentage depletion from Peters (1987, p. 257), and Part 8 is the net smelter return calculation from Peters (1987, p. 257).

14.2 USING THE MODEL

Using the model will demonstrate to you the outstanding benefit of 1-2-3 for a relatively complex worksheet. By entering new values in selected cells, recalculating, and graphing the results, you can determine quickly how changes affect a particular variable, such as DCFROR.

As an example of the procedure, I changed ore tonnage in place, D13, together with ore grade in place, D15..D17, using increments of plus and minus 10% and 20%. As tonnage increases, grade decreases, with the results graphed in Figures 14.1 and 14.2. Changing tonnage and grade in this way leads to a range in mine life between 10 and 16 years and a cash flow between about $4 and $8 million per year (Figure 14.1). DCFROR changes from 31% to 13%, illustrating that maximizing this measure depends on high cash flows early in the life of a project.

You can alter the model to suit yourself. Here are some examples of problems that may interest you:

1. Suppose that the ore tonnage in place, ORE.TONS.PLACE, changes inversely with the percentage extraction, %EXTRACTION, so that as the tonnage decreases the extraction rate increases.

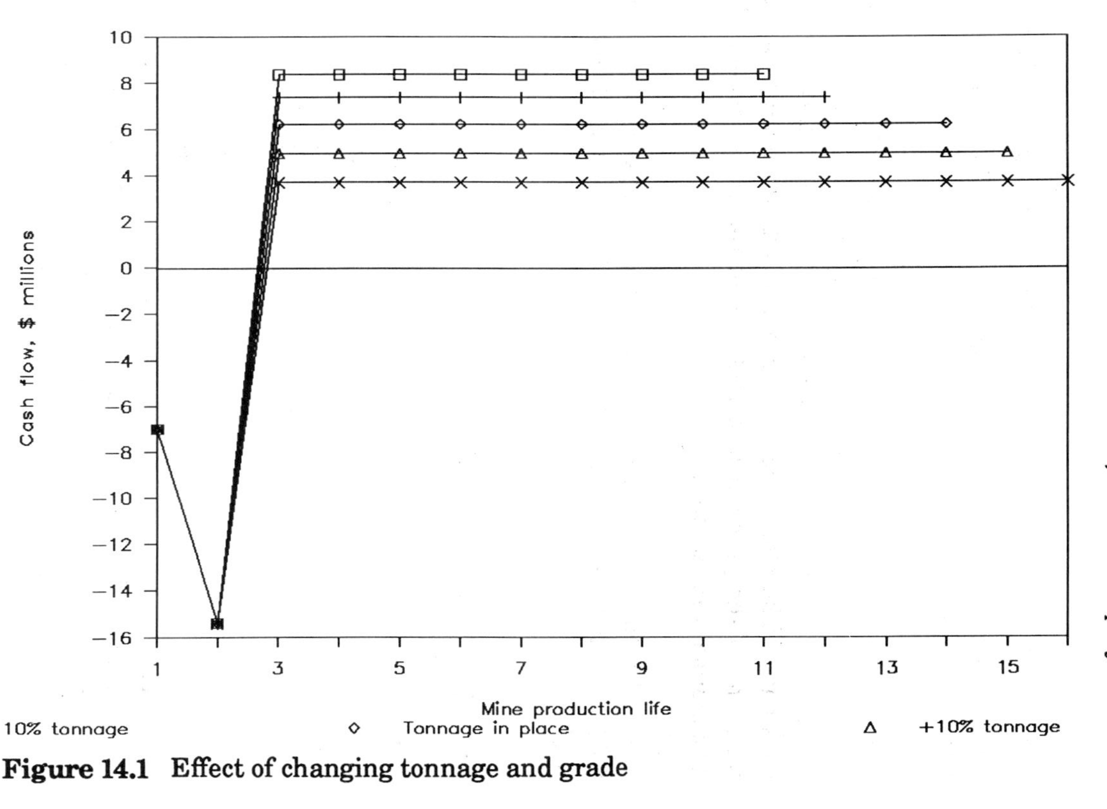

Figure 14.1 Effect of changing tonnage and grade

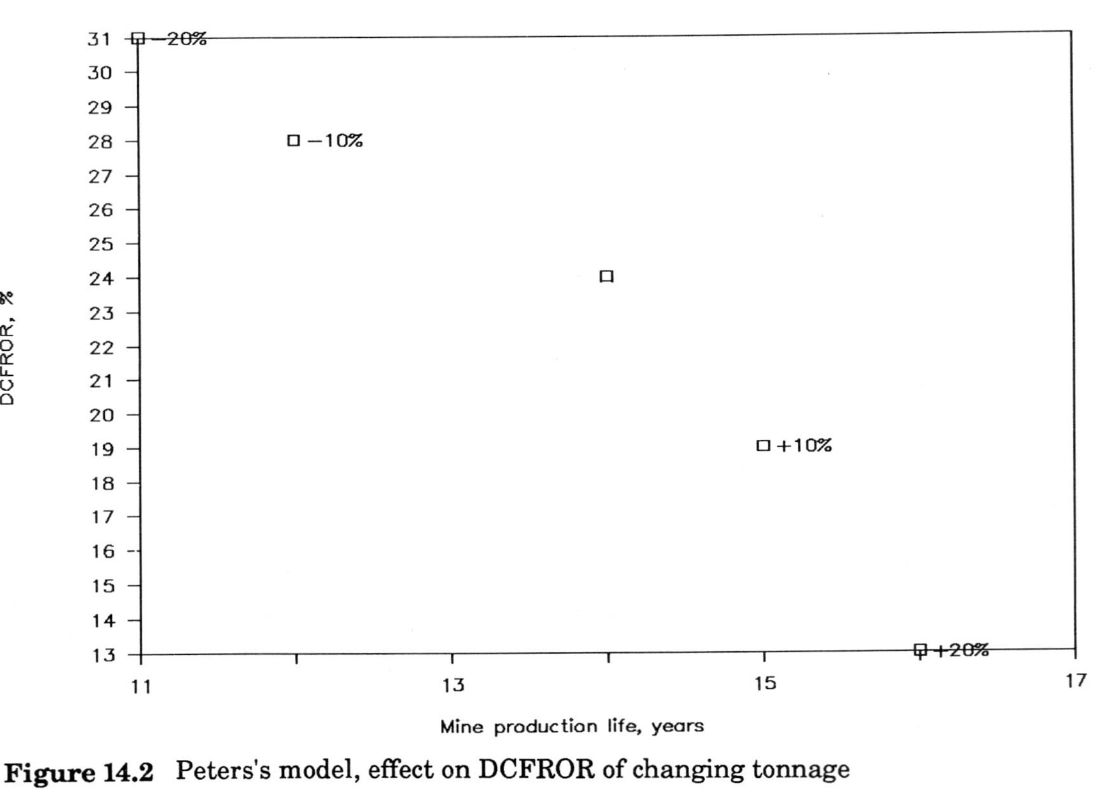

Figure 14.2 Peters's model, effect on DCFROR of changing tonnage

2. What happens if the dilution factor, %DILUTION, changes?

3. What happens if mill recovery, as indicated by the concentration ratios, PB.CONC.RATIO and ZN.CONC.RATIO, increases together with an increase in operating cost, OP.COST.PER.TON, required to achieve this improved mill recovery?

4. Suppose that tonnage, ORE.TONS.PLACE, changes along with mining rate, TONS.PER.DAY, so that the ratio of these two variables remains the same. Also, you may want to change the capital investment values—TOT.CAP.INVEST., DEPREC.YEAR.1, and DEPREC.YEAR.2 — assuming that a larger capital investment is needed for a higher mining rate.

5. Suppose that you change income tax, %INCOME.TAX, to correspond to situations where industry lobbyists are doing better or worse with the legislature.

6. Change revenue per ton broken ore, REV.PER.TON.ORE, accompanied by a change in operating costs, OP.COST.PER.TON, because the union has agreed to make wages change according to price.

7. Assume the same situation as in (6), except that the union president is difficult to work with so that wages and therefore operating costs, OP.COST.PER.TON, increase with lower revenue per ton, REV.PER.TON.ORE. At what point do you shut down the whole operation?

8. Assume that the plant is built at a lower price (changing TOT.CAP.INVEST., DEPREC.YEAR.1, and DEPREC.YEAR.2) but that the less expensive plant gets poorer mill recovery, as measured by PB.CONC.RATIO and ZN.CONC.RATIO.

Worksheet C14A Mineral property evaluation

```
 1  24 May 89, file C14A.WK1 on disks GSK 059-061
 2
 3  Worksheet C14A.WK1.  Mineral property evaluation
 4     (from Peters, 1987, p. 568-570)
 5
 6
 7  Part 1.  Summary data from working estimates
 8
 9                     C                          D
10   ------------------------------------------------------
11                      Item                     Example
12   ------------------------------------------------------
13   Ore tonnage in place, millions                  4.2
14   Ore grade in place, percent
15     Pb, percent                                  7.11
16     Zn, percent                                  5.64
17     Ag, g/T                                     155.4
18   Percentage extraction                          80.0
19   Dilution factor, percent                       15.0
20   Broken ore tonnage, millions                    3.9
21   Mining grade
22     Pb, percent                                  6.18
23     Zn, percent                                  4.90
24     Ag, g/T                                     135.1
25   Concentration ratio
26     Pb concentrate (100/PB%.CONC)               11.32
27     Zn concentrate (100/ZN%.CONC)               12.74
```

Worksheet C14A *(continued)*

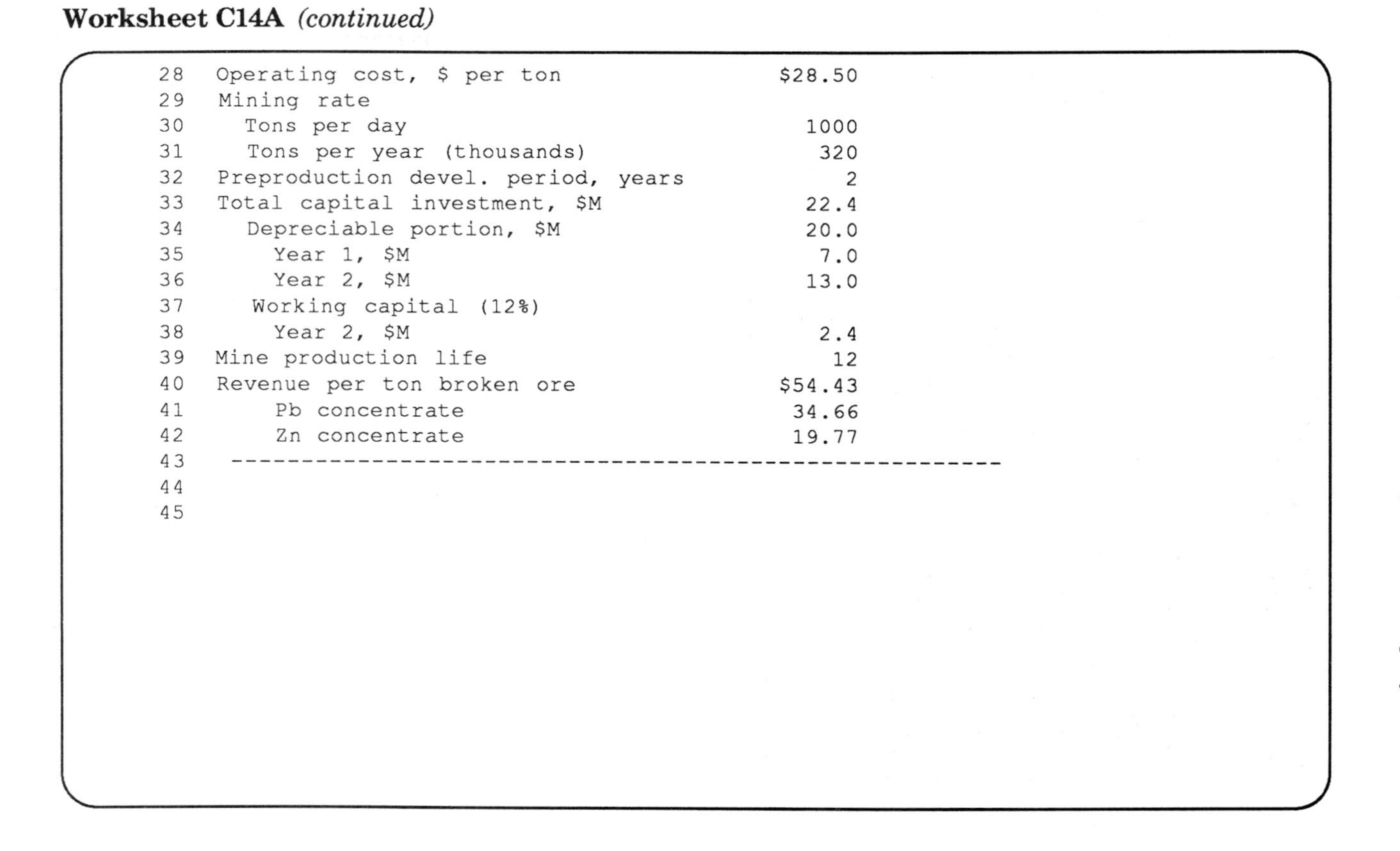

28	Operating cost, $ per ton	$28.50
29	Mining rate	
30	Tons per day	1000
31	Tons per year (thousands)	320
32	Preproduction devel. period, years	2
33	Total capital investment, $M	22.4
34	Depreciable portion, $M	20.0
35	Year 1, $M	7.0
36	Year 2, $M	13.0
37	Working capital (12%)	
38	Year 2, $M	2.4
39	Mine production life	12
40	Revenue per ton broken ore	$54.43
41	Pb concentrate	34.66
42	Zn concentrate	19.77
43	--	
44		
45		

Worksheet C14A *(continued)*

```
46  Part 2.  Fixed (or assigned) parameters
47
48                    C                          D
49   ------------------------------------------------------------
50                  Item                      Example
51   ------------------------------------------------------------
52   Equity 100% (unleveraged)
53   Constant dollars at project start
54  Royalty, percent                              3.0
55  Percentage depletion (in USA)                19.6
56     (See Appendix 3)
57  Federal and state income tax, %              51.0
58  Depreciation and amortization            straightline
59  Assigned hurdle rate for DCFROR, %           15.0
60   ------------------------------------------------------------
61
62
```

Worksheet C14A *(continued)*

63	Part 3. Annual estimate of cash flow ($, Millions)			
64				
65		D	E	F
66				
67			Year	
68				
69	Item	1	2	3-14
70				
71				
72	Revenue			$17.42
73	Royalty			0.52
74				
75	Gross income			16.90
76	Operating costs			9.12
77				
78	Operating income			7.78
79	Depreciation and amortization			1.67
80				
81	Net income before depletion			6.11
82				
83	Percentage depletion			3.31
84	50% limit			3.05
85	Selected depletion			3.05
86				
87	Taxable income			3.05
88	Income tax			1.56
89				

Worksheet C14A *(continued)*

```
 90  Net income                                                         1.50
 91  Add depreciation and amortization                                  1.67
 92  Add depletion                                                      3.05
 93
 94  Operating cash flow                                               $6.22
 95
 96  Capital expenditure                          -7.0     -13.0
 97  Working capital                                        -2.4
 98
 99  Cash flow                                   ($7.0)   ($15.4)      $6.22
100   ------------------------------------------------------------------------
101
102
103   Part 4.   Profitability expressions
104
105                     C                         D
106   -------------------------------------------------------------
107                    Item                    Example
108   -------------------------------------------------------------
109   Payback period, years                        3.6
110
111   DCFROR, percent                               24
112     Present value of income, $M             $15.7
113       P/A(I,N)                              3.871
114       P/F(I,2)                              0.652
115
116     Present value of capital invest.,$M     $15.7
117       P/F(I,1)                              0.807
118       P/F(I,2)                              0.652
```

Worksheet C14A *(continued)*

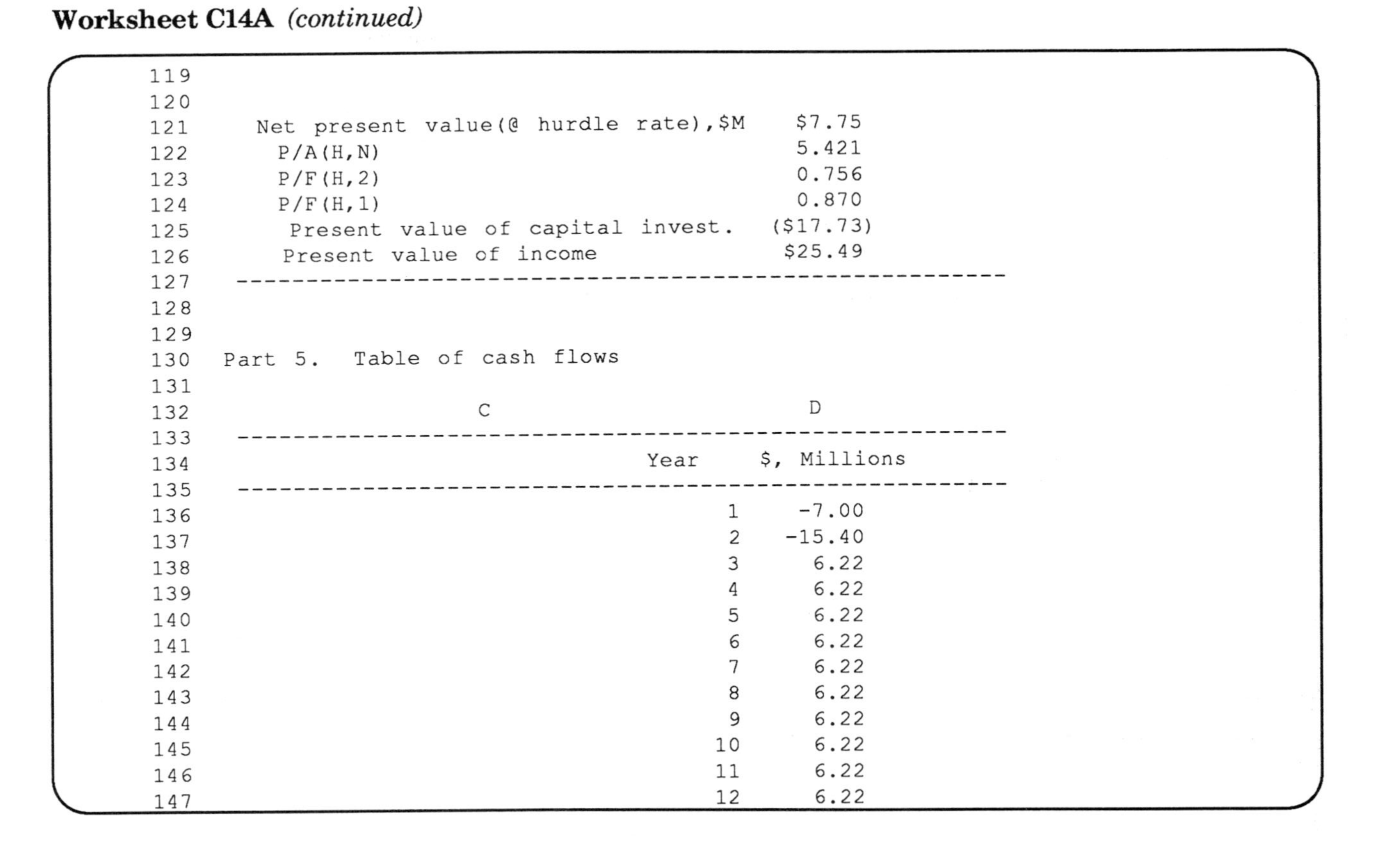

```
119
120
121    Net present value(@ hurdle rate),$M     $7.75
122      P/A(H,N)                               5.421
123      P/F(H,2)                               0.756
124      P/F(H,1)                               0.870
125       Present value of capital invest.   ($17.73)
126       Present value of income              $25.49
127   ----------------------------------------------------
128
129
130  Part 5.  Table of cash flows
131
132                    C                         D
133   ----------------------------------------------------
134                               Year    $, Millions
135   ----------------------------------------------------
136                                  1      -7.00
137                                  2     -15.40
138                                  3       6.22
139                                  4       6.22
140                                  5       6.22
141                                  6       6.22
142                                  7       6.22
143                                  8       6.22
144                                  9       6.22
145                                 10       6.22
146                                 11       6.22
147                                 12       6.22
```

Worksheet C14A *(continued)*

```
148                                          13      6.22
149                                          14      6.22
150                                          15      0.00
151                                          16      0.00
152                                          17      0.00
153                                          18      0.00
154                                          19      0.00
155                                          20      0.00
156   ------------------------------------------------------------
157
158
159  Part 6.  Detail of materials balance
160    (from Peters, 1987, p. 252)
161
162                  C                         D         E         F        G.
163   --------------------------------------------------------------------------------
164                                           Pb        Zn       Tail-     Heads
165                 Item                     Conc.     Conc.     ings
166   --------------------------------------------------------------------------------
167  Assays
168    Percent weight                          8.84      7.85     83.31    100.00
169    Ag, oz/mt                              44.57      3.13      0.19      4.35
170    Ag, g/mt                             1385.99     97.25      6.05     135.1
171    Pb, %                                  66.59      1.18      0.25      6.18
172    Zn, %                                   4.98     54.21      0.25      4.90
173
174  Contents
175    Ag, grams                                                  504.0   13513.0
```

Worksheet C14A *(continued)*

```
176    Ag, grams in Pb concentrate          12245.5
177    Ag, grams in Zn concentrate                      763.5
178    Pb, kilograms                          588.3       9.3       20.7     618.3
179    Zn, kilograms                           44.0     425.6       20.8     490.4
180
181   Percentage distribution
182    Ag                                     90.62      5.65       3.73    100.00
183    Pb                                     95.16      1.50       3.34    100.00
184    Zn                                      8.97     86.78       4.25    100.00
185   ----------------------------------------------------------------------------
186
187
188   Part 7.  Calculation of percentage depletion
189      (from Peters, 1987, p. 257)
190
191                   C                       D         E         F
192   ------------------------------------------------------------------
193                                          Pb        Zn
194                  Item                    Conc.     Conc.    Totals
195   ------------------------------------------------------------------
196   Concentrate selling price, $/ton      392.26    251.88
197   Ag, $                                  217.31      6.15    223.46
198   Pb, $                                 174.94
199   Zn, $                                            245.72
200   Pb + Zn, $                                                420.67
201   Ag, % of net smelter return (NSR)                              35
202   Pb + Zn, % of NSR                                              65
203   Percentage depletion from Ag             5.2
```

Worksheet C14A *(continued)*

```
204  Percentage depletion from Pb                  14.4
205  Percentage depletion                          19.6
206   ------------------------------------------------------------------
207
208
209  Part 8.  Net smelter return calculation
210     (from Peters, 1987, p. 257)
211
212                      C                          D
213   -----------------------------------------------
214                     Item                  Example
215   -----------------------------------------------
216  Metal prices
217    Silver
218      $ per troy ounce                        6.50
219      $ per gram                              0.21
220    Lead
221      $ per pound                             0.20
222      $ per kilogram                          0.44
223
224    Zinc
225      $ per pound                             0.45
226      $ per kilogram                          0.99
227
228  I.  Lead concentrate
229   -----------------------------------------------
230  Moisture, %                                    5
231  Freight charge, $/tonne                    10.00
232  Freight charge, $/dry tonne                10.53
```

Worksheet C14A *(continued)*

```
233
234  Payments                                 472.78
235    Silver                                 261.93
236    Lead                                   210.86
237
238   Deductions
239    Treatment charge                        70.00
240    Zn penalty                               0.00
241
242  Value at mine, per dry tonne             392.26
243
244
245   II.  Zinc concentrate
246  -------------------------------------------------
247  Moisture, %                                6.00
248  Freight charge, $/tonne                   12.00
249  Freight charge, $/dry tonne               12.77
250
251  Payments                                 420.64
252    Silver                                  10.27
253    Lead                                     0.00
254    Zinc                                   410.37
255
256   Deductions
257    Treatment charge                       156.00
258
259  Value at mine, per dry tonne             251.88
260
261
```

Worksheet C14A *(continued)*

```
262  III. Net smelter return (NSR) on ore
263  mined
264   -------------------------------------------------
265  NSR per tonne of ore                         54.43
266    Lead concentrates                          34.66
267    Zinc concentrates                          19.77
268   -------------------------------------------------
269
270
271  Part 9.  Table of labels
272
273  !CASH.FLOW.2                            E99
274  !CASH.FLOW.3                            F99
275  !DEPRE.&.AMORT.                         F79
276  !DPL.50%.LIMIT                          F84
277  !DPL.SELECTED                           F85
278  !GROSS.INCOME                           F75
279  !INCOME.TAX                             F88
280  !NET.BEFORE.DPL                         F81
281  !NET.INCOME                             F90
282  !OP.CASH.FLOW                           F94
283  !OP.INCOME                              F78
284  !OPERATING.COST                         F76
285  !PCT.DPL                                F83
286  !REVENUE                                F72
287  !ROYALTY                                F73
288  !TAX.INCOME                             F87
289  #P.A.H.N                                D122
290  #P.A.I.N                                D113
```

Worksheet C14A *(continued)*

```
291  #P.F.H.1            D124
292  #P.F.H.2            D123
293  #P.F.I.1            D117
294  #P.F.I.2            D114
295  %DCFROR             D111
296  %DEPLETION          D205
297  %DILUTION           D19
298  %EXTRACTION         D18
299  %HURDLE.RATE        D59
300  %INCOME.TAX         D57
301  %ROYALTY            D54
302  AG.!.PER.GRAM       D219
303  AG.G.IN.HEADS       G175
304  AG.G.IN.PB.CON      D176
305  AG.G.IN.TAILS       F175
306  AG.G.IN.ZN.CON      E177
307  AG.G.P.T.PB.CON     D170
308  AG.G.P.T.TAILS      F170
309  AG.G.P.T.ZN.CON     E170
310  AG.G.PER.T.ORE      D17
311  AG.ORE.MINED        D24
312  AG.PAY.PB.CON       D235
313  AG.PAY.ZN.CON       D252
314  AG.REC.PB.CON       D182
315  AG.REC.ZN.CON       E182
316  ANN.CASH.FLOWS       D136..D155
317  DEPR.CAP.INVEST     D34
318  DEPREC.YEAR.1       D35
319  DEPREC.YEAR.2       D36
```

Worksheet C14A *(continued)*

```
320  FREIGHT.PB.CON          D232
321  FREIGHT.ZN.CON          D249
322  HEADS.TONS              G168
323  LIFE                    D39
324  NSR.PB.CONC             D266
325  NSR.TOTAL               D265
326  NSR.ZN.CONC             D267
327  OP.COST.PER.TON         D28
328  ORE.TONS.BROKEN         D20
329  ORE.TONS.PLACE          D13
330  PB%.ORE                 D15
331  PB%.ORE.MINED           D22
332  PB.!.PER.KILO           D222
333  PB.%.IN.PB.CON          D171
334  PB.%.IN.TAILS           F178
335  PB.%.IN.ZN.CON          E171
336  PB.%.REC.PB.CON         D183
337  PB.%.REC.ZN.CON         E183
338  PB.CON.TREAT.           D239
339  PB.CONC.PRICE           D242
340  PB.CONC.RATIO           D26
341  PB.CONC.REVENUE         D41
342  PB.KG.IN.HEADS          G178
343  PB.KG.IN.PB.CON         D178
344  PB.KG.IN.ZN.CON         E178
345  PB.PAY.PB.CON           D236
346  PB.PAY.ZN.CON           D253
347  PB.WT.%.OF.CON          D168
348  PV.CAP.INVEST.          D125
```

Worksheet C14A *(continued)*

```
349  PV.INCOME               D126
350  REV.PER.TON.ORE         D40
351  TAILS.WT.%.CON          F168
352  TONS.PER.DAY            D30
353  TONS.PER.YEAR           D31
354  TOT.CAP.INVEST.         D33
355  TOT.PAY.PB.CON          D234
356  TOT.PAY.ZN.CON          D251
357  WORKING.CAPITAL         D38
358  ZN%.ORE                 D16
359  ZN%.ORE.MINED           D23
360  ZN.!.PER.KILO           D226
361  ZN.%.IN.PB.CON          D172
362  ZN.%.IN.ZN.CON          E172
363  ZN.%.REC.PB.CON         D184
364  ZN.%.REC.ZN.CON         E184
365  ZN.CON.TREAT.           D257
366  ZN.CONC.PRICE           D259
367  ZN.CONC.RATIO           D27
368  ZN.CONC.REVENUE         D42
369  ZN.KG.IN.HEADS          G179
370  ZN.KG.IN.PB.CON         D179
371  ZN.KG.IN.TAILS          F179
372  ZN.KG.IN.ZN.CON         E179
373  ZN.PAY.ZN.CON           D254
374  ZN.PENALTY              D240
375  ZN.WT.%.OF.CON          E168
```

References

Callahan, W.H., 1977, The history of the discovery of the zinc deposit at Elmwood, Tennessee, concepts and consequences: Econ. Geology, v. 72, no. 7, p. 1382-1392.

Cox, D.P., Wright, N.A., and Coakley, G.J., 1981, The nature and use of copper reserve and resource data: U.S. Geol. Survey Prof. Paper 907-F, p. F1-F20.

Davis, J.C., 1986, Statistics and data analysis in geology (2nd ed.): John Wiley & Sons, New York, 646 p.

Ewing, D.P., Noble, D.F., Burlakoff, G.S., Murray, K., Short, L.J., and Whitney, R., 1987, Using 1-2-3: Que Corporation, Indianapolis, Indiana, 905 p.

Ferguson, H.S., 1916, Placer deposits of the Manhattan district, Nevada: U.S. Geol. Survey Bull. 640-J, p. 163-193.

Fine, E.S., 1987, Plotting logarithmic graphs: Lotus [Magazine], March, 1987, p. 71-75.

Harbaugh, J.W., and Bonham-Carter, G., 1970, Computer simulation in geology: John Wiley & Sons, New York, 575 p.

Harbaugh, J.W., Doveton, J.H., and Davis, J.C., 1977, Probability models in oil exploration: John Wiley & Sons, New York, 269 p.

Harris, D.P., 1984, Mineral resources appraisal: Clarendon Press, Oxford, 445 p.

Hayes, W.B., and Koch, G.S., Jr., 1984, Constructing and analyzing area-of-influence polygons: Computers & Geosciences, v. 10, no. 4., p. 411-430.

Howarth, R.J., and Sinding-Larsen, 1983, Multivariate analysis, in Howarth, R.J., *ed.*, Handbook of exploration geochemistry, v.2, Statistics and data analysis in geochemical prospecting: Elsevier, Amsterdam, 207-289 p.

Kilpatrick, M., 1987, Business statistics using Lotus 1-2-3: John Wiley & Sons, New York, 326 p.

Koch, G.S., Jr., 1956, The Frisco mine, Chihuahua, Mexico: Econ. Geology, v. 51, no. 1, p. 1-40.

Koch, G.S., Jr., 1975, Unpublished report to Roan Consolidated Mines (Zambia) on consulting visit.

Koch, G. S., Jr., 1978, Ore estimation using confidence intervals and several weighting schemes — two practical examples: Paper presented at CENTO conference in Ankara, Turkey.

Koch, G.S., Jr., 1987, Exploration-geochemical data analysis with the IBM-PC: Van Nostrand Reinhold, New York, 175 p. (with 2 diskettes).

Koch, G.S., Jr., and R.F. Link, 1964, Accuracy in estimating metal content and tonnage of an ore body from diamond-drill-hole data: U.S. Bur. of Mines, Rept. of Invest. 6380, 24 p.

Koch, G. S., Jr., and R. F. Link, 1970, Statistical analysis of geological data, vol. 1: John Wiley & Sons, New York, 375 p.

Koch, G. S., Jr., and R. F. Link, 1971, Statistical analysis of geological data, vol. 2: John Wiley & Sons, New York, 438 p.

Koch, G.S., Jr., and Link, R.F., 1972, Sample preparation variability in diamond-drill core from the Homestake mine, Lead, S.D.: U.S. Bur. Mines, Rept. Invest. 7677, 14 p.

Koch, G.S., Jr., and Schuenemeyer, J.H., 1982, Exploration for zinc in Middle Tennessee by drilling: A statistical analysis: Econ. Geology, v. 77, no. 3, p. 653-663.

Koch, G.S., Jr., Schuenemeyer, J.H., and Link, R.F., 1974, A mathematical model to guide the discovery of ore bodies in a Coeur d'Alene lead-silver mine: U.S. Bur. Mines Rept. Invest. 7989, 43 p.

Lasky, S.J., 1950, How tonnage and grade relations help predict ore reserves: Engineering and Mining Journal, v. 151, no. 4, p. 81-85.

Li, J.C.R., 1964, Statistical inference I: Edwards Bros., Inc., Ann Arbor, Michigan, 658 p.

Mendelsohn, F., 1980, Some aspects of ore reserve estimation: University of the Witwatersrand, Johannesburg, South Africa, Economic Geology Research Unit, Inform. Circ. No. 147, 42 p.

Peters, W.C., 1987, Exploration and mining geology: John Wiley & Sons, New York, 685 p.

Ryan, B.F., Joiner, B.L., and Ryan, T.A., Jr., 1985, Minitab handbook, (2nd ed.): Duxbury Press, Boston, 374 p.

Sather, N.J., and Prindle, F.L., 1970, Milling practice at Bunker Hill, *in* Rausch, D.O, and Mariacher, B.C., eds., AIME world symposium on mining and metallurgy of lead and zinc: American Institute of Mining, Metallurgical, and Petroleum Engineers, New York, p. 348-372.

Smith, E.H., 1986, Metallurgy and mineral processing plant at St. Joe's El Indio mine in Chile: Mining Engineering, October, p. 971-979.

Stermole, F.J., and Stermole, J.M., 1987, Economic evaluation and investment decision methods: Investment Evaluations Corp., Golden, Colorado, 479 p.

Tukey, J.W., 1948, Approximate weights: Ann. Math. Statist., v. 19, p. 91-92.

Index